SOCIETY FOR EXPERIMENTAL BIOLOGY
SEMINAR SERIES · 17

CELLULAR ACCLIMATIZATON TO
ENVIRONMENTAL CHANGE

CELLULAR ACCLIMATISATION TO ENVIRONMENTAL CHANGE

Edited by

ANDREW R. COSSINS
Department of Zoology, University of Liverpool

PETER SHETERLINE
Department of Medical Cell Biology, University of Liverpool

CAMBRIDGE UNIVERSITY PRESS
Cambridge
London   New York   New Rochelle
Melbourne   Sydney

Published by the Press Syndicate of the University of Cambridge
The Pitt Building, Trumpington Street, Cambridge CB2 1RP
32 East 57th Street, New York, NY 10022, USA
296 Beaconsfield Parade, Middle Park, Melbourne 3206, Australia

© Cambridge University Press 1983

First published 1983

Printed in Great Britain by the University Press, Cambridge

Library of Congress catalogue card number: 82-17804

British Library Cataloguing in Publication Data
Cellular acclimatisation to environmental change.
– (Society for Experimental Biology seminar
series; 17)
1. Bioclimatology
I. Cossins, Andrew R. II. Sheterline, Peter
III. Series
591.52′22   QH543
ISBN 0 521 24384 X

UP

CONTENTS

List of contributors   vii

Introduction
A. R. Cossins and P. Sheterline   1

The adaptation of membrane structure and function to changes in temperature
A. R. Cossins   3

Mechanisms of homeoviscous adaptation in membranes
G. A. Thompson, Jr   33

Volume regulation by animal cells
E. K. Hoffmann   55

The use of *in vitro* enzyme activities to indicate the changes in metabolic pathways during acclimatisation
E. A. Newsholme and J. M. Paul   81

Cellular acclimatisation to environmental change by quantitative alterations in enzymes and organelles
B. D. Sidell   103

Cellular responses to an altered body temperature: the role of alterations in the expression of protein isoforms
I. A. Johnston   121

Membrane reorganisation and adaptation during chronic drug exposure
J. M. Littleton   145

Cellular adaptation of receptor-mediated function during opiate exposure
H. O. J. Collier   161

The induction of hepatic cytochrome(s) P_{450}: an adaptive response?
C. R. Elcombe   179

Cellular aspects of salinity adaptation in teleosts
J. C. Ellory and J. S. Gibson   197

Regulation of the seasonal biosynthesis of antifreeze peptides in cold-adapted fish
Y. Lin   217

Adaptation of erythrocyte function during changes in
environmental oxygen and temperature
D. A. Powers 227

Conclusions: a cellular perspective on environmental physiology
M. W. Smith, P. Sheterline and A. R. Cossins 245

Index 249

CONTRIBUTORS

Collier, H. O. J.
Miles Laboratories Ltd, Stoke Poges, Slough, Berks., UK. (Present address: Department of Pharmacology, Chelsea College, London SW3 6lX, UK.)

Cossins, A. R.
Department of Zoology, University of Liverpool, PO Box 147, Liverpool L69 3BX, UK.

Elcombe, C. R.
Central Toxicology Laboratories, ICI, Alderley Park, Macclesfield, Cheshire SK10 4TJ, UK.

Ellory, J. C.
Physiological Laboratory, University of Cambridge, Downing Street, Cambridge, UK.

Gibson, J. S.
Physiological Laboratory, University of Cambridge, Downing Street, Cambridge, UK.

Hoffmann, E. K.
Institute of Biological Chemistry A, University of Copenhagen, 13 Universitetparken, DK-2100, Copenhagen Ø, Denmark.

Johnston, I. A.
Department of Physiology, University of St Andrews, St Andrews, Fife, UK.

Lin, Y.
NCI-Frederick Cancer Research Centre, PO Box 13, Frederick, Maryland 21701, USA.

Littleton, J. M.
Department of Pharmacology, King's College, Strand, London WC2R 2LS, UK.

Newsholme, E. A.
Department of Biochemistry, University of Oxford, South Parks Road, Oxford, UK.

Paul, J. M.
Department of Biochemistry, University of Oxford, South Parks Road, Oxford, UK.

Powers, D. A.
Department of Biology, Johns Hopkins University, Baltimore, Maryland 21218, USA.

Sheterline, P.
Department of Medical Cell Biology, University of Liverpool, PO Box 147, Liverpool L69 3BX, UK.

Sidell, B. D.
Department of Zoology, University of Maine, Orono, Maine 04469, USA.

Smith, M. W.
ARC Institute of Animal Physiology, Babraham, Cambridge CB2 4AT, UK.

Thompson, Jr, G. A.
Department of Botany, University of Texas, Austin, Texas 78712, USA.

A. R. COSSINS and P. SHETERLINE

Introduction

Few or no organisms live in a constant natural environment such as that which can be provided experimentally for cells in culture. In view of this, it is not surprising that living organisms have evolved a rather versatile physiology which allows them to respond adaptively to the perturbations caused by changes in environmental conditions. For adaptations observed in natural 'field' conditions the term 'acclimatisation' is conventionally applied, whereas adaptations observed as a result of controlled conditions in the laboratory are termed 'acclimation'. This volume focuses on the cellular basis for acclimatisation (or acclimation) to a variety of environmental perturbations, including temperature, hypoxia, salinity and chemical perturbants, in the hope that both the scope and any common bases for these adaptive mechanisms may become clear.

Cellular mechanisms of acclimatisation are currently the subject of intense study. For the environmental physiologist these studies provide the basis for understanding the relationship of an organism with its surrounding, whilst for cell biologists they provide an important perspective on the degree of plasticity of cell structure and function and the adaptedness of the *status quo*. So whilst it is usually taken for granted that cells possess appropriate metabolic machinery, number and distribution of membrane-bounded compartments, etc., their dynamic nature and responsiveness to changes in environmental conditions are often less well appreciated. Thus it may be inappropriate to view structure and function of a given cell type as invariable; just as metabolic machinery has control systems to maintain an appropriate rate of product formation, it seems that the arrangement, composition and biological activity of other cellular structures may also be under adaptive control to maintain their effectiveness in changed conditions. Experimentally, one obvious means of demonstrating the existence of cellular control mechanisms and hence the dynamic nature of the cellular steady state is by perturbing the *status quo* and monitoring corrective responses. In this context, we have recognised two broad categories of adaptive response: first, those that maintain homeostasis of individual cells (dealt with in the first six articles of the volume), and secondly, adaptive responses of specialised cells which promote homeostasis of the whole organism (the last three articles of the volume). The progressive development of homeostatic properties in higher animals naturally increases the importance of the latter responses at the expense of the former, but also implies a shift in the level at which control is expressed, from intracellular to that mediated by nerves and hormones.

The presence of chemicals in the cellular environment, either self-administered (drugs) or from pollution may also severely perturb cellular function. Continued exposure to such chemicals may induce adaptive responses which offset their effects and may thus lead to tolerance and eventually to a state of dependence. This is considered in the articles by Littleton, Collier and Elcombe. It is clear that many of these responses are similar to those observed with other types of environmental insult and these studies hold out the exciting possibility that the cellular basis for drug tolerance, dependence and symptoms of withdrawal is a manifestation of cellular adaptation in general, which may thus be open to clinical manipulation.

A. R. COSSINS

The adaptation of membrane structure and function to changes in temperature

At the beginning of this century, Henriques & Hansen (1901) performed a rather bizarre experiment. They raised pigs in underwear at high temperatures and observed that the subcutaneous fats had a higher melting point. This was the first observation of a change in the properties of cellular lipids which tended to offset the perturbing effects of changed cellular temperature and thereby to preserve the physical state of those lipid compartments in some optimal or advantageous condition. More recently, this idea of adaptive change has found application to the adaptive responses of the cellular membranes of a wide variety of organisms, a process that has been termed 'homeoviscous adaptation' (Sinensky, 1974).

Since Henriques & Hansen's experiment, numerous studies have shown that reduced growth temperature or acclimation temperature in bacteria, plants and poikilothermic animals is invariably associated with the appearance of greater proportions of unsaturated fatty acids in membrane phospholipids (Hazel & Prosser, 1974). Because of their molecular geometry, phospholipids containing unsaturated fatty acids tend to have lower melting points, greater cross-sectional areas in the plane of the bilayer and a greater degree of molecular flexibility than their saturated homologues. The incorporation of greater proportions of unsaturated fatty acids in the cold, therefore, tends to offset the rigidifying and condensing effects of reduced temperature and hence to preserve a specific membrane condition. This hypothesis implies firstly the existence of an optimal membrane condition or fluidity for supporting efficient membrane function, and secondly that both fluidity and function are perturbed by temperature variations and are sensitive to variations in lipid composition.

It was not until the late 1960s and early 1970s with the development of various biophysical techniques to determine the phase state or degree of fluidity of biological membranes, that the hypothesis of homeoviscous adaptation of membranes could be tested, and a causal link established between altered membrane composition during thermal acclimation and an adaptation of membrane fluidity. In 1974 Sinensky showed, using electron spin resonance spectroscopy, that the membranes of *E. coli* were maintained in a roughly constant physical state at different growth temperatures. Subsequently, Nozawa *et al.* (1974) observed similar adaptive responses in the protozoan *Tetrahymena*, as did Cossins (1977) in the synaptosomal membranes of goldfish.

The purpose of this article is firstly to describe the evidence for the adjustment of membrane fluidity and phase state with variations in temperature, and secondly to demonstrate that this response does indeed influence the functional properties of membranes and thereby contribute to the overall cellular process of adaptation to environmental change.

Membrane dynamic structure and temperature

A thorough understanding of adaptive responses of membranes to temperature requires an appreciation of the direct effects of temperature upon membrane structure and function. Membranes of higher organisms appear to exist predominantly in the liquid-crystalline state at physiological temperatures (Quinn, 1981). This is characterised by a high degree of molecular motion within the bilayer. However, this liquid-like character is quite different from that of a simple, bulk, paraffinic solvent, since the hydrocarbon chains in each monolayer are aligned parallel to each other in a semi-crystalline array. This creates a distinctly anisotropic, semi-ordered environment. It is important to appreciate the dual nature of membrane structure, and that membrane order does not necessarily imply a lack of molecular motion.

Molecular motion may be of several distinct types, such as the flexing or twisting of hydrocarbon chains, the wobble and rotation of phospholipid molecules, the lateral diffusion of molecules along the plane of the membrane and the movement of molecules from one monolayer to the other. The resulting condition is generally described as 'fluid', although it should be appreciated that this is a general, conceptual term and due to the great complexity of motion in membranes has no specific, precise definition (Lands, 1980*b*). The available techniques for estimating the degree of membrane fluidity do not easily discriminate between the effects of order or of motion upon the spectroscopic properties of molecular probes. They usually provide information that relates to specific types of motion within the membrane which need not necessarily have relevance to other motional properties of membranes. Thus, nuclear magnetic resonance spectroscopy (NMR) provides detailed information on the motion of specific segments of hydrocarbon chains, whilst fluorescence polarisation reports on the motion of a molecular probe whose position within the bilayer is rather ill-defined (Cossins, 1981*a, b*). Unfortunately, it is not easy to apply NMR techniques to the problem of homeoviscous adaptation and most studies use fluorescence polarisation or electron spin resonance spectroscopy (ESR).

Additional complexity has been demonstrated by Seelig and his colleagues using NMR techniques. They have observed rather distinct structural zones, parallel to the membrane surface, that are created by the increasing mobility of the chains with depth into the bilayer, and by the presence of olefinic or unsaturated carbon–carbon bonds (Seelig & Seelig, 1980). Phospholipids with multiple olefinic bonds may give rise to a complex of structural zones which may be important for specific functional properties of membranes. Structural complexity of a different kind may also exist in the plane of the membrane due to fluid-phase immiscibility of different types of lipids (Wu & McConnell, 1975) or to the presence of a 'boundary' layer of lipids around protein molecules (Bennet, McGill & Warren, 1980). There is also evidence in certain membranes for a compositional asymmetry between the two leaflets of the bilayer which may lead to different fluidities in each monolayer (Rothman & Lenard, 1977).

An increase in temperature leads to a progressive increase in the fluidity of liquid-crystalline membranes, that is, an increase in molecular motion and a decrease in order. The temperature dependence of fluidity depends to a large extent upon the composition of the membrane; thus cholesterol reduces the Arrhenius activation energy of artificial membranes (Shinitzky & Inbar, 1976). Purple membranes of *Halobacterium* have a very low temperature dependence of fluidity (Kinosita *et al.*, 1981) which seems to be due to their unusually high protein to lipid ratios.

It is a fundamental property of phospholipid bilayers that they undergo changes in physical state at characteristic temperatures, from a liquid-crystalline state to a gel state in which the motion of each hydrocarbon chain is severely constrained by the crystalline alignment of its neighbours. In artificial membranes composed of defined phospholipids, phase transitions are highly co-operative and occur over a very narrow range of temperatures. Natural membranes contain highly complex mixtures of phospholipids as well as cholesterol and proteins, so that phase transitions may occur over a wide range of temperatures (Quinn, 1981). This means that at any temperature within the phase transition, patches of both liquid-crystalline and gel-phase lipids coexist, a condition known as a phase separation (Shimshick & McConnell, 1973). Phase transitions must pose very severe problems for organisms whose membranes usually exist in the liquid-crystalline state, and membrane functions and processes may be greatly perturbed. Membrane-bound enzymes undergo large increases in Arrhenius activation energy at temperatures below the phase transition, and some processes such as lateral diffusion may cease altogether (Melchior & Steim, 1976).

Homeoviscous adaptation

In Fig. 1(a) the fluidity of the liver mitochondrial membranes of 5 °C-acclimated and 25 °C-acclimated green sunfish are compared, to illustrate the general features of fluidity compensation. The important points to note are firstly that fluidity is temperature dependent and secondly that the graph for 5 °C-acclimated fish is shifted to lower temperatures compared with the corresponding graph for 25 °C-acclimated fish. Thus at any measurement temperature, the fluidity of membranes of cold-acclimated fish is somewhat greater than that of warm-acclimated fish. This obviously has the effect of offsetting the rigidifying effects of cold and is thus interpreted as a compensatory response whose adaptive significance lies in the preservation, to a greater or lesser extent, of a preferred membrane fluidity.

Fig. 1. (a) A comparison of the fluidity of liver mitochondrial fraction of 5 °C-acclimated (filled circles) and 25 °C-acclimated (open circles) green sunfish (*Lepomis cyanellus*). Fluidity was measured using the fluorescence polarisation technique with 1,6-diphenyl-1,3,5-hexatriene (DPH) as probe. The rotational diffusion coefficient ($\bar{R}$) was calculated according to the Perrin equation and used as an index of membrane fluidity. (From Cossins *et al.*, 1980.)

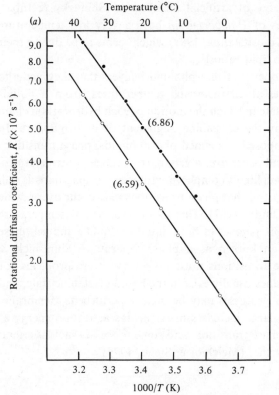

More recent time-resolved fluorescence studies have indicated that this compensation in fluidity of fish membranes during temperature acclimation is due more to a change in membrane order than to a change in the rate of rotational motion of the membrane constituents (Cossins, Kent & Prosser, 1980; Cossins & Prosser, 1982). This correlates well with the more expanded nature of phospholipids at lower acclimation temperatures (Haest, De Gier & Van Deenen, 1969; Cullen, Phillips & Shipley, 1971), which is due to the inclusion of more kinked unsaturation bonds which interfere with the close approach and potentially attractive interactions of adjacent hydrocarbon chains.

The main evidence for changes in the lipid composition of membranes being primarily responsible for homeoviscous responses, is that similar differences to those observed in the fluidity of natural membranes can be demonstrated in artificial membranes (liposomes) which have been reconstituted from the

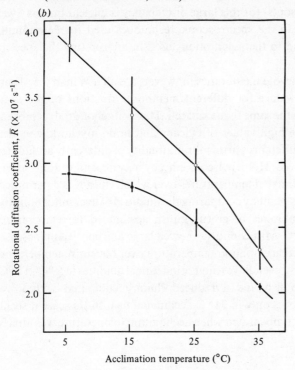

Fig. 1. (*b*) A comparison of homeoviscous adaptation of liver microsomal (filled circles) and mitochondrial (open circles) fractions of green sunfish acclimated to 5 °C, 15 °C, 25 °C and 34 °C. Fluidity was expressed as $\bar{R}$ at 25 °C. A horizontal line indicates no differences between membranes of differently acclimated fish. A negative slope indicates that membranes of cold-acclimated fish are more fluid than those of warm-acclimated fish. (From Cossins *et al.*, 1980.)

purified phospholipids of differently acclimated fish (Cossins, 1977) or *Tetrahymena* (Martin & Thompson, 1978; Thompson, 1980).

Homeoviscous efficacy

It is clear from Fig. 1(a) that the estimated fluidities of mitochondrial membranes of green sunfish were not identical at their respective acclimation temperatures, so that the homeoviscous response was somewhat less than complete or 'ideal' (Precht *et al.*, 1973). The relative magnitude of the observed compensatory response may be simply expressed as a fraction of that required for an 'ideal' response, and this has been termed 'homeoviscous efficacy' (Cossins, 1981a; Cossins & Prosser, 1982). The parallel nature of the graphs in Fig. 1(a), and incidentally in virtually all other observed homeoviscous responses, facilitates this measurement, since it is only necessary to determine the shift of the graph along the temperature axis as a result of acclimation and express it as a fraction of the difference in acclimation temperatures. A value of 1 indicates an 'ideal' response, whilst a value close to 0 indicates no response.

The values for various prokaryotic and eukaryotic organisms have been calculated from the published responses and are presented in Table 1. In prokaryotes, values between 0.1 and 1.0 have been observed even within the same species. The reason(s) for this large discrepancy is unclear but may well be due to differences in the spectroscopic techniques used for determining membrane fluidity and to the assumptions used in interpreting the spectroscopic data.

The values for eukaryote membranes, however, are all less than 0.5. There is some variation in values for different membrane fractions of the same species, and even from the same tissue and cell. Thus mitochondrial membrane fractions tend to have high values of efficacy and brain myelin low values. In some instances no differences in membrane fluidity of differently acclimated animals may be observed. Homeoviscous efficacy may depend to some extent upon the range of acclimated temperatures over which efficacy is determined. Fig. 1(b) illustrates this point by comparing the fluidities of liver mitochondrial and microsomal membranes of green sunfish acclimated to temperatures between 5 °C and 34 °C. Mitochondria showed large and consistent increases in fluidity with reduced acclimation temperature over the entire temperature range. Microsomes, however, have roughly indentical fluidities in 5 °C-, 15 °C- and 25 °C-acclimated fish and a reduced fluidity compared with lower acclimation temperatures only in 34 °C-acclimated fish. In this case it seems that fluidity was only compensated when acclimation temperature was altered to temperatures above 25 °C.

Table 1. *A comparison of homeoviscous efficacy in prokaryotes and eukaryotes*

Organism	Membrane fraction	Technique[a]	Range of growth or acclimation temp. (°C)	Efficacy[b]	Reference
E. coli	Outer	b	12–43	0.1–0.2	Janoff *et al.* (1979)
E. coli	Cytoplasmic	b	12–43	0.2	Janoff *et al.* (1979)
E. coli	Outer	a	20–37	0.25–0.4	Janoff *et al.* (1980)
Proteus mirabilis	Phospholipid extracts	d	15–43	1.0	Sinensky (1974)
		b	15–43	1.0	Rottem *et al.* (1978)
		c	15–43	0.3–0.5	Rottem *et al.* (1978)
B. stearothermophilus		b	42–50	1.0	Esser & Souza (1974)
Tetrahymena pyriformis	Cilia	b	15–34	0.25	Nozawa *et al.* (1974)
	Pellicle	b	15–34	0.2–0.5	Nozawa *et al.* (1974)
	Microsomes	b	15–34	0.25	Nozawa *et al.* (1974)
Carassius auratus (goldfish)	Synaptosomal	a	5–25	0.3	Cossins (1977)
	Sarcoplasmic reticulum	a	5–25	0.0	Cossins *et al.* (1978)
	Brain synaptic	a, e	7–28	0.36	Cossins & Prosser (1982)
	Brain mitochondrial	a, e	7–28	0.44	Cossins & Prosser (1982)
	Brain myelin	a, e	7–28	0.21	Cossins & Prosser (1982)
Lepomis cyanellus (green sunfish)	Liver mitochondrial	a	5–25	0.5	Cossins *et al.* (1980)
	Liver microsomal	—	5–25	0.3	Cossins *et al.* (1980)
Mesocricetus auratus (hamster, awake and hibernating)	Brain synaptosomal	a	4–37	0.0	Cossins & Wilkinson (1982)
	Kidney microsomal	a	4–37	0.0	Cossins & Wilkinson (1982)
Mesocricetus auratus	Brain microsomes	a	4–37	0.1	Goldman & Albers (1979)

[a] Techniques as follows: a, steady state fluorescence polarisation (1,6-diphenyl-1,3,5-hexatriene, DPH); b, electron spin resonance (ESR, 5-doxyl stearate); c, ESR (15-doxyl stearate or 16-doxyl stearate); d, ESR (methyl-12-doxyl stearate); e, differential polarised phase fluorimetry (DPH).
[b] Efficacy was calculated as the ratio of the shift of the fluidity/temperature graph along the temperature axis as a result of acclimation or hibernation, to the change in acclimation or body temperature. All values were calculated from data supplied in the original references.

Time-course of homeoviscous adaptation

In view of the desirability of establishing correlations with the time-course of lipid composition changes or functional modifications during thermal acclimation, there have been surprisingly few studies of the time-course of homeoviscous adaptation. In *Tetrahymena*, the changes in the fluidity of pellicle and microsomal membranes are essentially complete within 3–4 hours (Thompson, this volume). The time-course of the change in fluidity corresponded very closely with the time-course of changes in lipid composition and, incidentally, with changes in the onset of a phase transition (Martin & Thompson, 1978).

In the goldfish, Cossins, Friedlander & Prosser (1977) found that after transfer of 5 °C-acclimated goldfish to 25 °C it took approximately 10–15 days for the fluidity of brain synaptosomal membranes to reach a value that was characteristic of 25 °C-acclimated fish (Figure 2). During the reverse transfer, synaptosomal membrane fluidity remained almost constant for 20 days and then changed over the subsequent 20–30 days to a value similar to that for 5 °C-acclimated fish.

The reason for the wide discrepancy in the rates of homeoviscous adaptation of the goldfish and *Tetrahymena* is not clear but may be related to their respective rates of turnover of membrane components. Under normal culture

Fig. 2. The time-course of changes in membrane fluidity of goldfish brain synaptosomal fraction when 5 °C-acclimated fish were transferred to 25 °C and 25 °C-acclimated fish were transferred to 5 °C. Fluidity was measured as described in Fig. 1(*a*) and was expressed as fluorescence polarisation at 25 °C. A high value indicates low fluidity or high order, and vice versa. The dashed lines represent the approximate values of polarisation for membranes of animals fully acclimated to 5 °C (lower line) or to 25 °C (upper line). (Modified after Cossins *et al.*, 1977.)

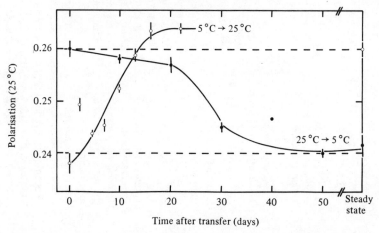

conditions *Tetrahymena* undergoes continuous and rapid cell division with the constant synthesis of cellular membranes. The neuronal population of higher organisms is, by comparison, rather static. Differences in turnover rate may also explain the differences in homeoviscous efficacy of different membrane fractions of a tissue, since fractions with a particularly slow rate of turnover, such as brain myelin, may not have achieved a steady state during the acclimation period commonly used in studies with fish (Cossins & Prosser, 1982).

Adaptations of membrane phase structure

Very drastic changes in the physical properties of membrane lipids, such as phase transitions, may be deleterious to the integrity and normal function of both cells and organisms (Raison, 1973; Melchior & Steim, 1976). In view of this, it is pertinent to ask whether the purpose of homeoviscous adaptation is to maintain an appropriate phase structure rather than to 'fine-tune' the fluidity of a liquid-crystalline membrane to some 'optimal' state. The adaptive benefits in the former case seem more dramatic and obvious because the disruption to normal function is greater, though this is no reason to exclude the evolution of a mechanism to fine-tune fluidity by small increments as a means of improving fitness. The main problem in answering this point, in eukaryotes at least, is our poor knowledge of the precise phase structure of membranes at physiological temperatures. In the case of fish, in which much is known concerning homeoviscous adaptation, there have been no determinations of lipid phase transition temperatures, though they are generally thought to be below 0 °C (Cossins, 1977, 1981*b*).

More is known about the phase structure of prokaryote membranes, and in *Acholeplasma*, at least, it seems that membrane fluidity *per se* need not be tightly regulated in order to preserve proper membrane function and support normal cellular activities (McElhaney, 1974; Silvius, Mak & McElhaney, 1980). Instead, homeoviscous adaptation serves to adjust membrane phase structure in a condition that permits normal function. The evidence in favour of adjustments of phase state when growth temperature is altered is strong. Sinensky (1974) found discontinuities in the Arrhenius plots of spin-probe motion which he interpreted as phase transitions. These occurred at approximately 15 deg C below growth temperature at all growth temperatures between 15 °C and 43 °C. Esser & Souza (1974) found similar discontinuities in *Bacillus stearothermophilus*, which occurred at their respective growth temperatures between 42 °C and 65 °C. A particularly good example was provided by Janoff, Haug & McGroarty (1979) who found two Arrhenius discontinuities in spin-probe motion in *E. coli* membranes, which they thought represented the upper and lower limits of a phase separation. Fig.

3(a) illustrates how the upper limit but not the lower limit varied with growth temperature, such that the phase separation was maintained at each growth temperature. In all three examples it appears that phase structure was maintained in a roughly constant state over wide ranges of growth temperature.

A similar conclusion also seems valid in the one example of the adaptation of phase structure in eukaryotes. Martin *et al.* (1976) have directly observed phase separations in *Tetrahymena* membranes by freeze-fracture electron microscopy. The creation of gel-phase lipids is thought to exclude protein molecules from the crystalline lattice and to lead to dense aggregations of proteins in the remaining fluid phase. Freeze-fracture electron microscopy reveals intramembranous particles which are generally thought to be due to the presence of intrinsic membrane proteins. Martin *et al.* (1976) used a measure of particle aggregation (particle density index) to determine the progress of phase separations in the alveolar membranes of *Tetrahymena* grown at 15 °C and 39 °C (Fig. 3b). At each growth temperature the particles appeared to be randomly dispersed, indicating the absence of a gel phase. Slight cooling, however, induced a progressive aggregation of particles. The shift of the particle density index/temperature curve along the temperature axis was 24 deg C, which corresponds exactly with the difference in growth temperatures, indicating a more precise control of phase transition temperatures than average fluidity (see Table 1).

Functional consequences of homeoviscous adaptation

Homeoviscous adaptation can only have truly adaptive significance through the modification of the functional properties of cellular membranes. Two important and related questions, therefore, are: 'which specific processes and functions are likely to be impaired by temperature variations?' and 'which functions are modified as a result of homeoviscous adaptation?' With respect to the first question, virtually all membrane processes and functions are affected by temperature variations, though it is difficult to decide which perturbations offer the greatest impairment to cellular activities. Silvius *et al.* (1980) have suggested that the barrier properties of membranes (i.e. permeability) may be particularly affected by large variations in fluidity and this could have very serious consequences for ion gradients across membranes.

The second question reduces to identifying those functions and processes of membranes that are sensitive to the magnitude of fluidity variations that occur during homeoviscous responses. In attempting to implicate homeoviscous adaptation as the mechanism of adaptive change in any specific case, it might be useful to itemise several criteria that in ideal conditions should be satisfied. Firstly, it is necessary to show that the process in question is

Fig. 3. (a) The alteration in upper (T_u) and lower (T_l) discontinuities of Arrhenius plots of spin-probe motion in membranes of *E. coli* grown at different temperatures. The stippled area indicates the temperature range over which the putative phase separation exists and the dashed line indicates growth temperature, T_g. (Data from Janoff *et al.*, 1979.)

(b) The alteration in phase separation temperatures of alveolar membranes of *Tetrahymena* when grown at 15 °C or 39 °C. The extent of the phase separation was estimated directly using freeze-fracture electron microscopy and the degree of aggregation of intramembrane particles was expressed as a particle density index. (Modified after Martin *et al.*, 1976.)

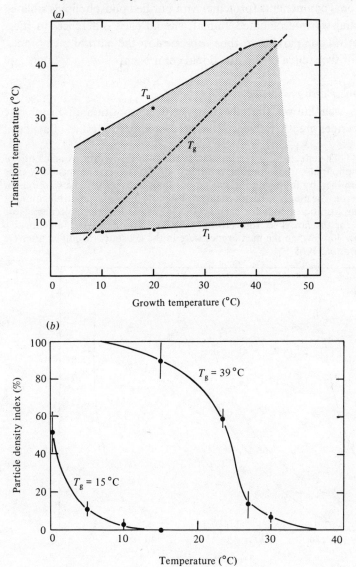

modified by experimental manipulations of membrane fluidity. A variety of procedures may be used in this respect, including lipid supplementation *in vivo* (dietary) or *in vitro*, delipidation with detergents or organic solvents followed by reconstitution with defined phospholipids, and, finally, the addition of drugs that fluidise the membrane. Secondly, it is necessary to demonstrate that both the fluidity and the function of the specific membrane type are altered during thermal acclimation. The direction of change, its extent and time-course should correspond in both instances. Finally, reconstitution of the functional components (proteins) with purified phospholipids isolated from cold- and warm-acclimated animals should elicit differences in functional properties comparable to those observed in the natural membranes, irrespective of the source of the functional components.

Permeability properties

It is well known that the fatty acid composition of membrane phosphoglycerides greatly influences the permeability properties of artificial

Fig. 4. The effects of increasing unsaturation of the hydrocarbon chains of phospholipids upon the permeability of liposomes. Permeability was determined by following the swelling of liposomes in isotonic solutions of glycerol. The numbers beside each graph denote the chain length and number of unsaturation bonds for each acyl chain of the lecithin molecules. Note that for distearoyl lecithin (18:0/18:0) the liposomes were impermeable below 40 °C when the membranes were in the gel state. (Modified after De Gier *et al.*, 1968.)

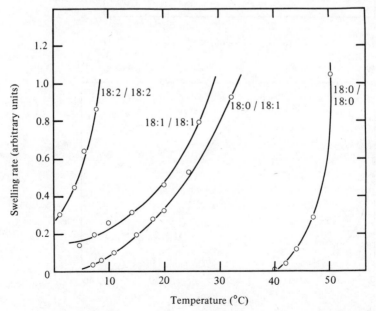

membranes to non-electrolytes (De Gier, Mandersloot & Van Deenen, 1968). Van Deenen and his colleagues have explored this relationship in some detail by following the swelling of liposomes when placed into an isosmotic solution of a permeable non-electrolyte. The rate of swelling is an index of permeability that may be conveniently estimated from the changes in light-scattering properties of the suspension. Fig. 4 shows that increased temperature leads to a large increase in permeability, and that the addition of olefinic bonds to the phospholipid leads to further increases in permeability. The effect of olefinic bonds upon permeability properties correlates closely with the increased cross-sectional area per phospholipid molecule in a monolayer and the increased fluidity of membranes containing unsaturated phospholipids (Haest et al., 1973).

Hazel (1979) has used similar techniques to compare the passive permeability of liposomes prepared from total liver phospholipid extracts of 5 °C- and 20 °C-acclimated rainbow trout (Fig. 5). He found that liposomes prepared from the phospholipids of cold-acclimated trout were more permeable than the corresponding liposomes of warm-acclimated trout. Indeed, the permeability coefficients measured at their respective acclimation temperatures were not dissimilar, illustrating just how large was this difference. Hazel & Schuster (1979) have subsequently shown that compensatory adjustments of permeability properties also occur in liver mitochondria of temperature-acclimated trout. Finally, Haest et al. (1973) found a close relationship between changes in phospholipid unsaturation, cross-sectional area per phospholipid molecule

Fig. 5. A comparison of the permeability to (a) erythritol and (b) glycerol of liposomes prepared from total liver phospholipid extracts of 5 °C- and 20 °C-acclimated rainbow trout. (Modified after Hazel, 1979.)

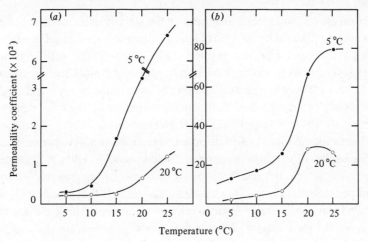

and permeability properties of liposomes when *E. coli* was cultured at different temperatures.

These experiments show that the changes in passive permeability properties with temperature acclimation conform to many of the criteria discussed previously and there seems little doubt that the restructuring of membranes during acclimation results in adaptive changes in passive permeability properties. The precise physiological significance of basal permeability and its adaptation is, however, questionable. Recent work has revealed that what were previously regarded as 'leaks' or basal permeabilities of membranes are, in fact, dominated by facilitated diffusion mechanisms (Hoffmann, this volume). In addition, the discovery of a class of pore-forming proteins in bacteria and in mitochondria (Zalman, Nikaido & Kagawa, 1980) provides another means by which membrane permeability may be enhanced well above that of a phospholipid bilayer. Both facilitated diffusion mechanisms and pore-forming antibiotics are barely influenced by lipid phase transitions or by changes in membrane fluidity (Read & McElhaney, 1976; Boheim, Hanke & Hansjörg, 1980), so these processes would seem to be little affected by homeoviscous adaptation.

The activity of membrane-bound enzymes

A principal feature of the fluid-mosaic model of membrane structure is the close relationship between the phospholipid bilayer and intrinsic membrane proteins. Since proteins are apparently responsible for much of the biological activity of cellular membranes, such as ion transport, lipid biosynthesis, oxidative phosphorylation and cell-surface phenomena, it is not surprising that a great deal of attention has been paid to their properties and to the influence of the surrounding bilayer on their structural and kinetic properties.

Studies on the effect of membrane order on the activity of membrane-bound enzymes are complicated by the idea of layer(s) of semi-immobilised phospholipid which surround intrinsic proteins. The evidence for this so-called boundary layer comes from the restriction of spin-probe mobility in reconstituted systems with very low phospholipid to protein ratios, when it is thought that the existence of this less fluid microcompartment may be detected (Griffith *et al.*, 1973; Vanderkooi, 1974; Hesketh *et al.*, 1976). Chapman, Gomez-Fernandez & Goni (1979) dispute this concept and have proposed an alternative explanation which relies on restricted probe mobility in phospholipid domains which become trapped within protein aggregates at low phospholipid to protein ratio. Nevertheless, it is possible that boundary layer lipids may form a physically distinct microenvironment which buffers the protein from the bulk bilayer. Thus techniques which measure the average fluidity of membranes, such as ESR or fluorescence spectroscopy, need not necessarily provide information that is relevant to the effective viscous environment of the protein.

Delipidation and reconstitution. Phospholipids are required for the enzymatic activity of a large number of membrane-bound proteins (Coleman, 1973) and procedures which remove the phospholipid environment of such proteins cause partial or complete inactivation. In most cases the lipid specificity of membrane-bound enzymes is broad (Bennet *et al.*, 1980; Hokin, 1981). The acyl chain composition of the phospholipids, however, may influence the level of activity and this fact suggests that the restructuring of membrane lipids during thermal acclimation may modulate enzymatic activity in an adaptive manner.

This hypothesis has been supported by the studies of Hazel (1972) on succinate dehydrogenase of the epaxial muscle of goldfish. Extraction of soluble succinate dehydrogenase from goldfish muscle mitochondria caused a reduction, but not a complete loss, of enzymatic activity. Regardless of the source of the delipidated enzyme, its activity was restored to a higher level by reconstitution with a total mitochondrial lipid extract from 5 °C-acclimated goldfish than with a lipid extract from 25 °C-acclimated goldfish. Hazel showed that the magnitude of the reactivation was dependent mainly on the unsaturation of the phospholipid acyl chains.

Fluidity and enzymatic activity. A particularly elegant demonstration of the modulation of enzymatic activity by the physical state of the membrane has been provided by Sinensky *et al.* (1979). These workers have isolated a mutant of a Chinese hamster ovary cell line that was defective in the regulation of cholesterol biosynthesis. Cultures were grown under conditions in which plasma membranes were enriched with various cholesterol contents which produced membranes with different membrane fluidities. The specific activity of the $(Na^+ + K^+)$-ATPase from these membranes was shown to vary in an exponential manner with an order parameter that was measured by ESR spectroscopy (Fig. 6). The number of ATPase molecules was not affected by the cholesterol supplementation procedures, so the conclusion that the turnover number was modulated by the degree of order of the membrane seems justified and certainly supports studies with reconstituted systems (Kimelberg & Papahadjopoulos, 1974).

This effect can be explained by assuming that enzymes, in general, undergo conformational changes during catalysis and that this transition is rate-limiting (Cleland, 1975). The work required to produce a conformational change in the polypeptide chain is largely made up of work required to produce the complementary conformational changes in the neighbouring hydrocarbon chains, and Sinensky et al. (1979) point out that this work is linearly related to the order parameter. Bearing in mind that the effective solvent environment of intrinsic enzymes is highly viscous and distinctly anisotropic compared with aqueous media, then the restriction of protein conformational flexibility by the lipid bilayer fluidity may be especially pronounced.

Recent studies on certain aqueous enzymes provide strong evidence that proteins are not rigid structures but are highly dynamic and that solvent viscosity is the dominant rate-limiting factor in both conformational transitions (Beece et al., 1980) and enzymatic activity (Gavish & Werber, 1979). It is thus clear that the molecular flexibility of proteins endows them with a sensitivity to solvent viscosity and that, in principle, the manipulation of membrane fluidity by various means, including homeoviscous adaptation, can have marked effects upon the rates of enzyme catalysis and hence the functional properties of membranes.

Fig. 6. The dependence of the specific activity of the $(Na^+ + K^+)$-ATPase of Chinese hamster ovary cell plasma membranes, upon lipid acyl chain order. Different membrane orders were produced by cholesterol supplementation of a mutant that was defective in cholesterol synthesis. (After Sinensky et al., 1979.)

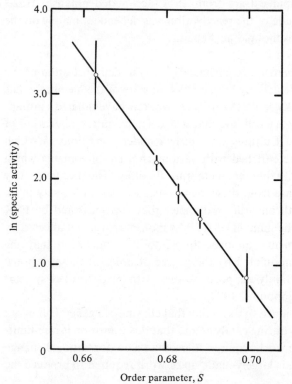

That the specific activity of membrane-bound enzymes is altered during thermal acclimation has been shown several times, though whether this is due to altered turnover or to altered number of catalytic units is not always resolved. One exception is the $(Na^+ + K^+)$-ATPase of the intestinal mucosa of goldfish, which showed a fall in specific activity with increased acclimation temperature (Smith & Ellory, 1971). Studies with radiolabelled ouabain, a specific inhibitor of this enzyme, demonstrated that the concentration of the enzyme was not significantly altered by acclimation, so that the calculated turnover number fell by 30% when 16 °C-acclimated goldfish were transferred to 30 °C. More recently, Wodtke (1981) has shown significant increases in the turnover number of several enzymes of the electron transport chain in liver mitochondria of cold-acclimated carp relative to warm-acclimated carp.

In some cases the activity of membrane-bound enzymes is inversely related to membrane fluidity. Riordan, Alon & Buchwald (1977) found that the Mg^{2+}-ATPase of rat liver plasma membranes exhibited a negative temperature dependence above 31 °C. Manipulation of membranes by altering their phospholipid composition or cholesterol content resulted in enhanced activity in more ordered membranes (Kimelberg, 1975; Riordan, 1980). This unusual phenomenon was explained by invoking the 'vectorial displacement hypothesis' of Shinitzky and his colleagues (Borochov & Shinitzky, 1976), in which the protein may be vertically displaced by changes in membrane order, such that its active site(s) becomes more exposed at the surface of the membrane to enable greater enzymatic activity.

Since the precise function of the Mg^{2+}-ATPase is unknown, the physiological significance of these observations is unclear, but these findings do suggest that homeoviscous adaptation may modulate enzymes in an entirely different way to that described for the $(Na^+ + K^+)$-ATPase. Indeed, Smith & Ellory (1971) noted increased activity of the Mg^{2+}-ATPase of mucosal homogenates of goldfish intestine during warm-acclimation. Also, Lagerspetz & Skytta (1979) have shown a higher Mg^{2+}-ATPase activity in epidermal homogenates of 25 °C-acclimated frogs relative to 6 °C-acclimated frogs. Following Duncan (1967) these authors suggest that this enzyme controls some aspect of the passive permeability of epidermal cells and that the increased activity in the warm accounts for the perfect temperature compensation of Na^+ transport across frog skin (Lagerspetz & Skytta, 1979).

Fluidity and protein dynamics. A somewhat different approach to the effects of homeoviscous adaptation upon membrane-bound proteins has been adopted by Cossins, Bowler & Prosser (1981). They compared the stability at high inactivating temperatures of the synaptic $(Na^+ + K^+)$-ATPase of cold- and warm-acclimated goldfish, in the belief that the process of inactivation, in common with normal catalytic activity, is a manifestation of the conformational freedom or flexibility of the constituent polypeptide chains in the enzyme. Both processes involve conformational transitions which one might reasonably expect to be rate-limited by the thermal motion of the surrounding bilayer. Thus, an enzyme that was more highly constrained or restricted by its solvent environment would be inactivated at higher temperatures than a less constrained enzyme.

Fig. 7 shows an experiment where synaptic membrane preparations were pre-incubated for 15 minutes over a range of inactivating temperatures and the activity remaining was subsequently assayed at a non-inactivating temperature. The enzyme from cold-acclimated fish was inactivated at lower temperatures than the enzyme of warm-acclimated fish. This difference is quite large and the rate constants for inactivation were two-fold greater in the former (R. N. A. H. Lewis, K. Bowler & A. R. Cossins, unpublished observations).

These differences in thermal stability could be due, in principle, either to a modification of the enzyme by interactions with specific modulators, to the induction of temperature-specific isoenzymes, or to the viscotropic influence of an altered membrane fluidity. Support for the last alternative comes from the demonstration of a marked decrease in thermal stability of the enzyme when the membrane was fluidised by the anaesthetic n-hexanol (Cossins *et*

Fig. 7. A comparison of the thermal stability of the $(Na^+ + K^+)$-ATPase of 6 °C-acclimated (open symbols) and 28 °C-acclimated (filled symbols) goldfish. Membrane preparations were pre-incubated at different temperatures for 15 minutes and the residual enzymatic activity was assayed at 25 °C. (After Cossins *et al.*, 1981.)

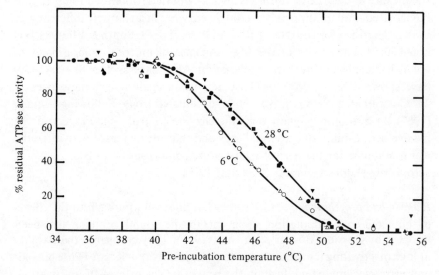

al., 1981). The main conclusion from these studies was that the conformational freedom and molecular flexibility of the $(Na^+ + K^+)$-ATPase are heavily influenced by changes in membrane fluidity that occur during temperature acclimation. Although these experiments relate to a non-physiological process, it seems reasonable to extend the effects of fluidity adaptations to include the modulation of protein flexibility and hence enzyme turnover rate at physiological temperatures.

Arrhenius discontinuities and phase changes. A sizeable literature exists to show that membrane-bound enzymes may display sharp discontinuities in the Arrhenius plots of enzymatic activity (Raison, 1973). In some cases the temperatures of these discontinuities correspond with calorimetrically determined phase transitions (Kruyff *et al.*, 1973) or with discontinuities of spin-probe motion (see Vignais & Devaux, 1976). In other cases the discontinuity is abolished by solubilisation of enzymes with detergents. These effects have usually been interpreted as demonstrations of the dramatic influence of membrane phase state on the activation energy of the reaction and, as such, discontinuities may be used as a means of studying the phase properties of membranes. While this may be true in some instances, it is by no means applicable to all enzymes with curvilinear or discontinuous Arrhenius plots (see Cossins, 1981*b*). A notable exception was found by Dean & Tanford (1979), who observed a discontinuity of the Ca^{2+}-ATPase of muscle sarcoplasmic reticulum even when freed from endogenous lipids by detergent solubilisation.

Wodtke (1976) has found convincing discontinuities in the Arrhenius plots of succinate oxidation by liver mitochondria of carp. The temperature of the discontinuity was dependent upon acclimation temperature: 14–16 °C for 10 °C-acclimated fish and 22–24 °C for 26 °C-acclimated fish. Furthermore, Wodtke has demonstrated a progressive decrease over 15–25 days in the discontinuity temperature of 26 °C-acclimated carp when fish were transferred to and maintained at 10 °C. Whilst in the absence of supporting biophysical information it is premature to attribute these discontinuities to lipid phase transitions, the behaviour of this system is at least consistent with the idea that the apparently adaptive changes in this mitochondrial process are related to homeoviscous adaptation.

Houslay & Palmer (1978) have provided convincing evidence of an adjustment of phase behaviour of the plasma membranes of hamster liver during hibernation. The adenyl cyclase of these membranes is thought to exist predominantly within the inner monolayer of the membrane, whilst the hormone receptor with which it may interact is situated on the external monolayer. Basal adenyl cyclase activity in normothermic hamsters exhibited an Arrhenius discontinuity at 26 °C, but glucagon-stimulated adenyl cyclase activity showed discontinuities at 26 °C and 13 °C. Houslay & Palmer (1978) suggested that these discontinuities were respectively due to phase transitions at 26 °C in the inner monolayer and 13 °C in the outer monolayer, and this interpretation was supported by the observation of similar discontinuities for other membrane-bound enzymes which were also thought to be asymmetrically positioned within the membrane. In hibernating hamsters the Arrhenius discontinuities were found at 26 °C and 4 °C for basal and hormone-stimulated activities, respectively. This naturally leads to the conclusion that the phase behaviour of the outer monolayer only is altered during hibernation. The adaptive value of a response which prevents a potentially deleterious phase transition is obvious, though why this should apply only to the outer monolayer of this membrane is not so clear. However, this study does provide the first evidence for an asymmetry of adaptive response between the two monolayers of a membrane.

Transport processes

We have seen how phospholipids play a crucial role in supporting the enzymatic activity of some membrane-bound enzymes. Recent advances in the techniques of reconstitution of transport proteins into phospholipid bilayers have not only provided unequivocal proof that these proteins are responsible for specific transport processes but also that phospholipids must form sealed structures, that is, create an effective diffusion barrier across which transport functions may be expressed (Hokin, 1981).

Perhaps the best evidence for the modulation of transport activity by membrane fluidity comes from studies with an unsaturated fatty acid auxotroph of *E. coli* which show rather distinct breaks in the Arrhenius plots of both glucose and galactose transport. The temperatures of these breaks responded to changes in membrane fatty acid composition in a manner that was consistent with their being related to lipid phase transitions (Schairer and Overath 1969; Wilson, Rose & Fox, 1970). Thus cells grown in a medium supplemented with oleate (18:1) exhibited a break at 14 °C, whilst cells grown in the presence of linolenate (18:2) showed a break at 7 °C. Linden *et al.* (1973) subsequently showed that a second Arrhenius discontinuity could be detected if studies were extended to higher temperatures, and that these discontinuities correlated with changes in the slope of Arrhenius plots of the solubility of a spin-probe in the membrane. Similar correlations between lipid composition and transport activity have been observed in eukaryote cells by Kaduce *et al.* (1977).

The preceding discussion concerning the compensation of $(Na^+ + K^+)$-ATPase activity during temperature acclimation suggests that the transport function of this enzyme should also be altered in an apparently adaptive manner. Possibly the best-characterised epithelial transport system from this point of view is the intestinal mucosa of goldfish. Smith (1970) demonstrated that both Na^+ transport and fluid transport across the intestine were much reduced, at constant incubation temperature, by acclimation of goldfish to high temperatures. Smith & Ellory (1971) showed that these changes occurred between 12 and 36 hours after transfer from 16 °C to 30 °C, there being no subsequent change for up to 20 days at 30 °C. The activity of the $(Na^+ + K^+)$-ATPase, however, showed no similar changes during the first 2 days after transfer and only changed slowly to a lower level over the following 20 days. This discrepancy in the time-course of changes in these two processes indicates that some other process is subject to adaptive control, probably the diffusional or carrier-mediated entry of Na^+ at the mucosal surface (Smith, 1976). This study also demonstrates the need to compare the time-courses of adaptive changes in order to establish correlations.

Lagerspetz & Skytta (1979) have observed 'ideal' or 'perfect' compensations of Na^+ transport across frog skin. Thus the short-circuit current measured with an Ussing chamber was identical at each acclimation temperature. Moreover, this compensation was totally reversible on reacclimation. Very little change was seen in the activity of the $(Na^+ + K^+)$-ATPase in epidermal homogenates. The Mg^{2+}-ATPase showed a decrease in activity at lower acclimation temperature and these authors also believe that this enzyme is involved in the control of passive permeability properties of the epidermal cell at the mucosal surface.

Bourne & Cossins (1981) have looked for adaptive changes during acclimation of carp, in the active transport in that most popular of transport systems, the erythrocyte. Although there were distinct differences in the temperature dependence of active and passive K^+ influx, they were opposite to what one might have expected. There was, however, some evidence of increased thermal lability in the cells of cold-acclimated fish. The unidirectional fluxes at the steady state are, of course, dependent upon the respective concentrations of the co-transported cations and ATP. Variations in any of these factors with acclimation may contribute to the observed differences, and in future studies they must be taken into account. It is clear from all of these studies that the adaptation of enzymatic activity does not necessarily mean that the associated active transport process at the normal steady state is also adapted in any obvious way. It seems that other contributory processes which are rate-limiting to the overall transport process may be subject to adaptive control and that these obscure the observation of adaptations in a specific active transport mechanism.

Conclusions

Eukaryotic cells have the ability to create a set of specific and differentiated membrane types (Morré, Kartenbeck & Franke, 1979). Each has a distinctive and characteristic molecular composition (Van Deenen, 1965) with which to perform its appointed tasks and each has a specific fluidity which presumably suits those functional properties. The precise nature of the optimal condition in any specific instance is not known though it is likely to reflect a compromise between the diverse structural requirements for the various functional properties of the membrane. In broad terms, it may be a compromise between the rate-depressing effects of membrane order on matrix processes, such as enzymatic activity or lateral diffusion, and the rate-enhancing effects of membrane disorder on barrier properties.

The idea of an appropriate or optimal membrane fluidity is reinforced by the demonstration of homeoviscous responses, in which adjustments in fluidity appear to compensate for the membrane perturbation. Homeoviscous adaptation can be viewed as the manifestation of a control mechanism that forms part of the normal repertoire of cells even under steady-state conditions, rather than as a special mechanism which operates to offset temperature effects. Thus the control mechanism that normally operates to create a set of differentiated and adapted membranes is revealed by the existence of compensatory responses to membrane perturbations.

There is a good deal of evidence that these adaptive responses are mediated by changes in the unsaturation of membrane phosphoglycerides (Thompson, this volume) and this hypothesis is now approaching the status of dogma. However, it is worth emphasising that most techniques for estimating membrane fluidity are not very specific with regard to the location of the probe or the distribution of microdomains within the membrane. The adaptation of fluidity may well result from a shift in the distribution of fluid and gel-phase microdomains in a phase separation, or a shift in the distribution of the probe between these microdomains, or between the lipid bilayer and proteins. Thus a change in the phospholipid to cholesterol ratio or the phospholipid to protein ratio may also be involved in homeoviscous adaptation, although there is not a great deal of evidence to this effect.

The extent of homeoviscous adaptation varies between different membrane types, even within the same cell or tissue, and in some instances does not occur. Homeoviscous adaptation is not, therefore, a ubiquitous response of living organisms. Arguments against the adaptive benefits of homeoviscous responses can be made in those instances where no fluidity compensation occurs (Cossins, Christiansen & Prosser, 1978; Cossins & Wilkinson, 1982) and one can speculate that it is apparent only when some selective benefit accrues. Present information suggests that, at best, the process is able to offset only half of the direct effects of temperature upon membrane fluidity. However, there is no reason to believe that maintaining a constant fluidity at each acclimation temperature is necessarily the most appropriate response, and the terms 'ideal' or 'complete' may impose false objectives and limits for the process.

Homeoviscous adaptation has some of the characteristics of homeostatic systems in general and it is instructive to interpret the responses in terms of a control system. Membrane fluidity is a function of membrane composition and this is controlled by the activity of biosynthetic processes. The central question is how the biosynthetic processes are modulated to produce an adaptive shift in membrane fluidity. Temperature may have a direct effect upon biosynthetic processes, which are 'programmed' to create the appropriate adaptive response (Fig. 8a). Thus temperature acts as a stimulus to an open control system (see Wilson (1979) for a summary of control processes).

Alternatively, membrane fluidity may itself influence the biosynthetic processes in such a way as to offset the perturbation (Fig. 8b). In this system, temperature acts as a disturbance to the controlled variable (fluidity), and a negative feedback loop produces the appropriate adjustment in biosynthesis. The biosynthetic apparatus would, therefore, act both as the sensor of the controlled variable and as the effector of the response. This scheme has some

Fig. 8. Schematic diagrams to illustrate an open (a) and a closed (b) feedback control system which may control membrane lipid composition and membrane fluidity.

(a) Open control system

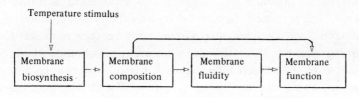

(b) Closed feedback loop

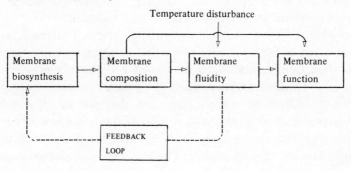

attractive features. Firstly, the response is flexible and may be closely adjusted to the magnitude and the direction of the perturbation. Secondly, it suggests a mechanism for the control of lipid biosynthesis, namely a viscotropic effect of fluidity on both the rates and specificity of membrane biosynthesis. Thus a membrane that is more ordered than the 'optimal' condition would induce the biosynthesis of a greater proportion of phospholipids containing unsaturated fatty acids and vice versa. However, this description raises the question as to how the biosynthetic apparatus recognises the 'optimal' condition (i.e. the set point) by which the scale and direction of the perturbation can be assessed and the appropriate response induced. Either one implicates an additional controlling influence or the proteins of the biosynthetic apparatus have been constructed in such a way as to respond in the appropriate manner. Thompson (1980) has suggested that vertical movements of biosynthetic enzymes in the membrane may be induced by variations in the order of the membrane in such a way as to modify the specificity or activity of fatty acid desaturases. Melchior & Steim (1977), on the other hand, suggest that it may be unnecessary to invoke enzymatic mechanisms for the selective incorporation of certain fatty acids, rather that the differential solubility of saturated and unsaturated fatty acids in the membrane may provide the required fluidity-programmed selectivity. Indeed, Lands (1980a) maintains that at present there is no evidence for a direct 'dialogue' between membrane fluidity and biosynthesis.

An important difference between the closed and open control systems described here, is that factors other than temperature which perturb membrane fluidity should elicit homeoviscous responses in the former but not the latter. Relevant perturbations are hydrostatic pressure, fluidising drugs and altered lipid diet. An open system must be 'programmed' for each perturbing influence separately and since many membrane systems are not naturally exposed to such influences it is difficult to envisage how such a system can be selected for during evolution. The closed, feedback system responds directly to variations in the controlled variable (fluidity) no matter how these changes are produced. The case for homeoviscous responses to drugs is considered in detail by Littleton (this volume).

The adaptive value of the homeoviscous response depends upon the compensation of the functional properties of membranes for variations in temperature. Considerable evidence is available to show that many membrane functions and processes are dependent, to a greater or lesser extent, upon bulk membrane fluidity. In principle, therefore, they will be affected both by temperature variations and by homeoviscous adaptation. The sheer complexity and extensive microheterogeneity of membranes makes it difficult to predict which process will be most influenced by homeoviscous adaptation since they may be sequestered in non-adapting or poorly adapting microenvironments. Nevertheless, compensatory changes in the enzymatic activity and stability of some membrane-bound proteins, and in the permeability properties of artificial and natural membranes, as a result of thermal acclimation have been demonstrated which are consistent with their dependence upon homeoviscous adaptation. The crucial and unanswered question is whether these compensations of performance are vital either to the survival of individuals in a crisis, or to the overall fitness of a species in a seasonally changing environment.

Perhaps the most convincing evidence of a link between homeoviscous adaptation and the functional compensation of membranes comes from the demonstration of adaptive effects in isolated functional systems when reconstituted with the purified phospholipids of differently acclimated organisms. So far, this has been achieved by Hazel's (1972) study of the mitochondrial enzyme, succinate dehydrogenase, and by Hazel & Schuster's (1979) demonstration of adaptation of liposomal permeability. No doubt as the techniques for the reconstitution of membrane proteins improve this will become an important area of study.

So far as we know, the strategy of modifying the effective viscosity of membranes for adaptive purposes is one that applies only to the hydrophobic compartment and is made possible by the unusual nature of this compartment. The potentially large number of properties and processes of membranes which may be influenced by homeoviscous adaptation makes it a very potent adaptive mechanism. However, it is certainly not the only adaptive process that operates during thermal acclimation. Variations in the concentrations of membrane-bound enzymes and in the cellular distribution and arrangement of membranes (Penny & Goldspink, 1980; Sidell, this volume) may act in conjunction with homeoviscous adaptation to influence specific processes to a greater extent than can be achieved simply by variations in membrane fluidity.

References

Beece, D., Eisenstein, H., Frauenfelder, H., Good, D., Marden, M. C., Reinisch, L., Reynolds, A. H., Sorensen, L. B. & Yue, K. T. (1980). Solvent viscosity and protein dynamics. *Biochemistry*, **19**, 5147–57.

Bennet, J. P., McGill, K. A. & Warren, G. B. (1980). The role of lipids in the functioning of a membrane protein: the sarcoplasmic reticulum calcium pump. In *Current Topics in Membranes and Transport*, Vol. 14, ed. F. Bronner & A. Kleinzeller, pp. 128–64. New York & London: Academic Press.

Boheim, G., Hanke, W. & Hansjörg, E. (1980). Lipid phase transition in planar bilayer membrane and its effect on carrier and pore-mediated ion transport. *Proceedings of the National Academy of Sciences, USA*, **77**, 3403–7.

Borochov, H. & Shinitzky, M. (1976). Vertical displacement of membrane proteins mediated by changes in microviscosity. *Proceedings of the National Academy of Sciences, USA*, **73**, 4526–30.

Bourne, P. K. & Cossins, A. R. (1981). The effects of thermal acclimation upon ion transport in erythrocytes. *Journal of Thermal Biology*, **6**, 179–81.

Chapman, D., Gomez-Fernandez, J. C. & Goni, F. M. (1979). Intrinsic protein–lipid interactions; physical and biochemical evidence. *FEBS Letters*, **98**, 211–23.

Cleland, W. W. (1975). What limits the rate of an enzyme-catalysed reaction? *Accounts of Chemical Research*, **8**, 145–51.

Coleman, R. (1973). Membrane-bound enzymes and membrane ultrastructure. *Biochimica et Biophysica Acta*, **300**, 1–30.

Cossins, A. R. (1977). Adaptation of biological membranes to temperature. The effect of temperature acclimation of goldfish upon the viscosity of synaptosomal membranes. *Biochimica et Biophysica Acta*, **470**, 395–411.

Cossins, A. R. (1981a). The adaptation of membrane dynamic structure to temperature. In *Effects of Low Temperature on Biological Membranes*, ed. G. J. Morris & A. Clarke, pp. 82–106. New York & London: Academic Press.

Cossins, A. R. (1981b). Steady state and dynamic fluorescence studies of the adaptation of cellular membranes to temperature. In *Fluorescent Probes*, ed. G. Beddard & M. A. West, pp. 39–80. New York & London: Academic Press.

Cossins, A. R., Bowler, K. & Prosser, C. L. (1981). Homeoviscous adaptation and its effects upon membrane-bound enzymes. *Journal of Thermal Biology*, **6**, 183–7.

Cossins, A. R., Christiansen, J. & Prosser, C. L. (1978). Adaptation of biological membranes to temperature. The lack of homeoviscous adaptation in the sarcoplasmic reticulum. *Biochimica et Biophysica Acta*, **511**, 442–54.

Cossins, A. R., Friedlander, M. J. & Prosser, C. L. (1977). Correlations between behavioural temperature adaptations of goldfish and the viscosity and fatty acid composition of their synaptic membranes. *Journal of Comparative Physiology*, **120**, 109–21.

Cossins, A. R., Kent, J. & Prosser, C. L. (1980). A steady state and differential polarised phase fluorimetric study of the liver microsomal and mitochondrial membranes of the thermally-acclimated green sunfish (*Lepomis cyanellus*). *Biochimica et Biophysica Acta*, **599**, 341–58.

Cossins, A. R. & Prosser, C. L. (1982). Variable homeoviscous responses of different brain membranes of thermally-acclimated goldfish. *Biochimica et Biophysica Acta*, **687**, 303–9.

Cossins, A. R. & Wilkinson, H. (1982). The role of homeoviscous adaptation in mammalian hibernation. *Journal of Thermal Biology*, **7**, 107–10.

Cullen, J., Phillips, M. C. & Shipley, G. G. (1971). The effects of temperature on the composition and physical properties of the lipids of *Pseudomonas fluorescens*. *Biochemical Journal*, **125**, 733–42.

Dean, W. L. & Tanford, C. (1978). Properties of a delipidated, detergent-activated Ca^{2+}-ATPase. *Biochemistry*, **17**, 1683–90.

De Gier, J., Mandersloot, J. G. & Van Deenen, L. L. M. (1968). Lipid composition and permeability of liposomes. *Biochimica et Biophysica Acta*, **150**, 666–75.

Duncan, C. J. (1967). *The Molecular Properties and Evolution of Excitable Cells*. Oxford: Pergamon Press.

Esser, A. F. & Souza, K. A. (1974). Correlation between thermal death and membrane fluidity in *Bacillus stearothermophilus*. *Proceedings of the National Academy of Sciences, USA*, **71**, 4111–15.

Gavish, B. & Werber, M. M. (1979). Viscosity-dependent structural fluctuations in enzyme catalysis. *Biochemistry*, **18**, 1269–75.

Goldman, S. S. & Albers, R. W. (1979). Cold resistance of the brain during hibernation: changes in the microviscosity of the membrane and associated lipids. *Journal of Neurochemistry*, **32**, 1139–42.

Griffith, O. H., Jost, P. C., Capaldi, R. A. & Vanderkooi, G. (1973). Boundary lipid and fluid bilayer regions in cytochrome oxidase model membranes. *Annals of the New York Academy of Sciences*, **222**, 561–73.

Haest, C. W. M., De Gier, J. & Van Deenen, L. L. M. (1969). Changes in the chemical and barrier properties of the membrane lipids of *E. coli* by variation of the temperature of growth. *Chemistry and Physics of Lipids*, **3**, 413–17.

Hazel, J. R. (1972). The effect of temperature acclimation upon succinic dehydrogenase activity from the epaxial muscle of the common goldfish (*Carassius auratus* L.). *Comparative Biochemistry and Physiology*, **43B**, 863–82.

Hazel, J. R. (1979). Influence of thermal acclimation on membrane lipid composition of rainbow trout liver. *American Journal of Physiology*, **236**, R91–R101.

Hazel, J. R. & Prosser, C. L. (1974). Molecular mechanisms of temperature compensation in poikilotherms. *Physiological Reviews*, **54**, 620–77.

Hazel, J. R. & Schuster, V. L. (1979). The effects of temperature and thermal acclimation upon the osmotic properties and non-electrolyte permeability of liver and gill mitochondria from rainbow trout (*Salmo gairdneri*). *Journal of Experimental Zoology*, **195**, 425–38.

Henriques, V. & Hansen, C. (1901). Vergleichende Untersuchungen über die chemische Zusammensetzung des thierischen Fettes. *Skandinavia Archives für Physiologie*, **11**, 151–65.

Hesketh, T. R., Smith, G. A., Houslay, M. D., McGill, K. A., Birdsall, N. J. M., Metcalfe, J. M. & Warren, G. B. (1976). Annular lipids determine the ATPase activity of a calcium transport protein complexed with dipalmitoyl lecithin. *Biochemistry*, **15**, 4145–51.

Hokin, L. E. (1981). Reconstitution of 'carriers' in artificial membranes. *Journal of Membrane Biology*, **60**, 77–93.

Houslay, M. D. & Palmer, R. W. (1978). Changes in the form of Arrhenius plots of the activity of glucagon-stimulated adenylate cyclase and other hamster liver plasma-membrane enzymes occurring on hibernation. *Biochemical Journal*, **174**, 909–19.

Janoff, A. S., Gupte, S. & McGroarty, E. J. (1980). Correlation between temperature range of growth and structural transitions in membranes and lipids of *E. coli*. *Biochimica et Biophysica Acta*, **598**, 641–4.

Janoff, A. S., Haug, A. & McGroarty, E. J. (1979). Relationship of growth temperature and thermotropic phase changes in cytoplasmic and outer membranes from *E. coli* K12. *Biochimica et Biophysica Acta*, **555**, 56–66.

Kaduce, T. L., Awad, A. B., Fontanelle, L. J. & Spector, A. A. (1977). Effect of fatty acid unsaturation on α-amino isobutyric acid transport in Ehrlich ascites cells. *Journal of Biological Chemistry*, **252**, 6624–30.

Kimelberg, H. K. (1975). Alterations in phospholipid-dependent ($Na^+ + K^+$)-ATPase activity due to lipid fluidity. Effects of cholesterol and Mg^{2+}. *Biochimica et Biophysica Acta*, **413**, 143–56.

Kimelberg, H. K. & Papahadjopoulos, D. (1974). Effects of phospholipid acyl chain fluidity, phase transitions and cholesterol on $(Na^+ + K^+)$-stimulated ATPase. *Journal of Biological Chemistry*, **249**, 1071–80.

Kinosita, K. Jr, Kataoka, R., Kimura, Y., Gotoh, O. & Ikegami, A. (1981). Dynamic structure of biological membranes as probed by 1,6-diphenyl-1,3,5-hextriene: a nanosecond fluorescence depolarisation study. *Biochemistry*, **20**, 4270–7.

Kruyff, B. De, Van Dijck, P. W. M., Goldbach, R. W., Demel, R. A. & Van Deenen, L. L. M. (1973). Influence of fatty acid and sterol composition on the lipid phase transition and activity of membrane-bound enzymes in *Acholeplasma laidlawii*. *Biochimica et Biophysica Acta*, **330**, 269–82.

Lagerspetz, K. Y. H. & Skytta, M. (1979). Temperature compensation of sodium transport and ATPase in frog skin. *Acta Physiologica Scandinavica*, **106**, 151–8.

Lands, W. E. M. (1980a). Dialogue between membranes and their lipid-metabolising enzymes. *Transactions of the Biochemical Society*, **8**, 25–7.

Lands, W. E. M. (1980b). Fluidity of membrane lipids. In *Membrane Fluidity: Biophysical Techniques and Cellular Regulation*, ed. M. Kates & A. Kuksis, pp. 69–73. Clifton, New Jersey: Humana Press.

Linden, C. D., Wright, K. L., McConnel, H. M. & Fox, C. F. (1973). Lateral phase separations in membrane lipids and the mechanism of sugar transport in *Escherichia coli*. *Proceedings of the National Academy of Sciences, USA*, **70**, 2271–2275.

McElhaney, R. N. (1974). The effect of membrane-lipid phase transitions on membrane structure and on the growth of *Acholeplasma laidlawii*. *Journal of Supramolecular Structure*, **2**, 617–20.

Martin, C. E., Hiramitsu, K., Kitajima, Y., Nozawa, Y., Skriver, L. & Thompson, G. A. (1976). Molecular control of membrane properties during temperature acclimation. Fatty acid desaturase regulation of membrane fluidity in acclimating *Tetrahymena* cells. *Biochemistry*, **15**, 5218–27.

Martin, C. E. & Thompson, G. A. Jr (1978). Use of fluorescence polarisation to monitor intracellular membrane changes during temperature acclimation. Correlation with lipid compositional and ultrastructural changes. *Biochemistry*, **17**, 3581–6.

Melchior, D. L. & Steim, J. M. (1976). Thermotropic transitions in biomembranes. *Annual Reviews of Biophysics and Bioengineering*, **5**, 205–38.

Melchior, D. L. & Steim, J. M. (1977). Control of fatty acid composition of *Acholeplasma laidlawii* membranes. *Biochimica et Biophysica Acta*, **466**, 148–59.

Morré, D. J., Kartenbeck, J. & Franke, W. W. (1979). Membrane flow and interconversion among endomembranes. *Biochimica et Biophysica Acta*, **559**, 71–152.

Nozawa, Y., Iida, H., Fukushima, H. & Ohnishi, S. (1974). Studies on *Tetrahymena* membranes: temperature-induced alterations in fatty acid composition of various membrane fractions in *Tetrahymena pyriformis* and its effect on membrane fluidity as inferred by spin-label study. *Biochimica et Biophysica Acta*, **367**, 134–47.

Penny, R. K. & Goldspink, G. (1980). Temperature adaptation of sarcoplasmic reticulum of fish muscle. *Journal of Thermal Biology*, **5**, 63–8.

Precht, H., Christophersen, J., Hensel, H. & Larcher, W. (1973). *Temperature and Life*, pp. 334–7. Berlin: Springer Verlag.

Quinn, P. J. (1981). The fluidity of cell membranes and its regulation. *Progress in Biophysics and Molecular Biology*, **38**, 1–104.

Raison, J. K. (1973). The influence of temperature-induced phase changes on the kinetics of respiratory and other membrane-associated enzymes. *Journal of Bioenergetics*, **4**, 285–309.

Read, B. D. & McElhaney, R. N. (1976). Influence of membrane lipid fluidity on glucose and uridine facilitated diffusion in human erythrocytes. *Biochimica et Biophysica Acta*, **419**, 331–41.

Riordan, J. R. (1980). Ordering of bulk membrane lipid or protein promotes activity of plasma membrane Mg^{2+}ATPase. *Canadian Journal of Biochemistry*, **58**, 928–34.

Riordan, J. R., Alon, N. & Buchwald, M. (1977). Plasma membrane lipids of human diploid fibroblasts from normal individuals and patients with cystic fibrosis. *Biochimica et Biophysica Acta*, **574**, 39–47.

Rothman, J. E. & Lenard, J. (1977). Membrane asymmetry. *Science*, **195**, 743–53.

Rottem, S., Markowitz, O. & Razin, S. (1978). Thermal regulation of the fatty acid composition of lipopolysaccharides and phospholipids of *Proteus mirabilis*. *European Journal of Biochemistry*, **85**, 455–50.

Schairer, H. U. & Overath, P. (1969). Lipids containing *trans*-unsaturated fatty acids change the temperature characteristics of thiodimethylgalactoside accumulation in *Escherichia coli*. *Journal of Molecular Biology*, **44**, 209–14.

Seelig, J. & Seelig, A. (1980). Lipid conformation in model membranes and biological membranes. *Quarterly Reviews of Biophysics*, **13**, 19–61.

Shimshick, E. J. & McConnell, H. M. (1973). Lateral phase separation in phospholipid membranes. *Biochemistry*, **12**, 2351–60.

Shinitzky, M. & Inbar, M. (1976). Microviscosity parameters and protein mobility in biological membranes. *Biochimica et Biophysica Acta*, **433**, 133–49.

Silvius, J. R., Mak, N. & McElhaney, R. N. (1980). Why do prokaryotes regulate membrane fluidity? In *Membrane Fluidity: Biophysical Techniques and Cellular Regulation*, ed. M. Kates & A. A. Kuksis, pp. 213–22. Clifton, New Jersey: Humana Press.

Sinensky, M. (1974). Homeoviscous adaptation – a homeostatic process that regulates the viscosity of membrane lipids in *E. coli*. *Proceedings of the National Academy of Sciences, USA*, **71**, 522–5.

Sinensky, M., Pinkerton, F., Sutherland, E. & Simon, F. R. (1979). Rate limitation of $(Na^+ + K^+)$-stimulated ATPase by membrane acyl chain ordering. *Proceedings of the National Academy of Sciences, USA*, **76**, 4893–7.

Smith, M. W. (1970). Selective regulation of amino acid transport by the intestine of goldfish (*Carassius auratus* L.). *Comparative Biochemistry and Physiology*, **35**, 387–401.

Smith, M. W. (1976). Temperature adaptation in fish. *Biochemical Society Symposia*, **41**, 43–60.

Smith, M. W. & Ellory, J. C. (1971). Temperature-induced changes in sodium transport and Na^+/K^+-ATPase activity in the intestine of goldfish (*Carassius auratus* L.). *Comparative Biochemistry and Physiology*, **39A**, 209–18.

Thompson, G. A. Jr (1980). Regulation of membrane fluidity during temperature acclimation by *Tetrahymena pyriformis*. In *Membrane*

Fluidity: Biophysical Techniques and Cellular Regulation, ed. M. Kates & A. A. Kuksis, pp. 381–97. Clifton, New Jersey: Humana Press.

Van Deenen, L. L. M. (1965). Permeability and topography in membranes. In *Progress in the Chemistry of Fats and Related Lipids*, vol. 8, part 1, ed. R. Holman, pp. 1–27. Oxford: Pergamon Press.

Vanderkooi, G. (1974). Organisation of proteins in membranes with special reference to the cytochrome oxidase system. *Biochimica et Biophysica Acta*, **344**, 307–45.

Vignais, P. M. & Devaux, P. F. (1976). The use of spin labels to study membrane-bound enzymes, receptors and transport systems. In *The Enzymes of Biological Membranes*, vol. 1, ed. A. N. Martonosi, pp. 91–117. New York: Wiley.

Wilson, G., Rose, S. P. & Fox, C. P. (1970). The effect of membrane lipid unsaturation upon glycoside transport. *Biochemical and Biophysical Research Communications*, **38**, 617–23.

Wilson, J. A. (1979). *Principles of Animal Physiology*, 2nd edn. New York: Macmillan.

Wodtke, K. (1976). Discontinuities in the Arrhenius plots of mitochondrial membrane-bound enzyme systems from a poikilotherm: acclimation temperature of carp affects transition temperatures. *Journal of Comparative Physiology*, **110**, 145–57.

Wodtke, E. (1981). Temperature adaptation of biological membranes. Compensation of the molar activity of cytochrome c oxidase in the mitochondrial energy-transducing membrane during thermal acclimation of the carp (*Cyprinus carpio* L.). *Biochimica et Biophysica Acta*, **640**, 710–20.

Wu, S. H. & McConnell, H. M. (1975). Phase separations in phospholipid membranes. *Biochemistry*, **14**, 847–54.

Zalman, L. S., Nikaido, H. & Kagawa, Y. (1980). Mitochondrial outer membrane contains a protein producing non-specific diffusion channel. *Journal of Biological Chemistry*, **255**, 1771–4.

GUY A. THOMPSON, JR

Mechanisms of homeoviscous adaptation in membranes

Among the many strategies used by cells during acclimation to environmental stress, only one clearly involves a definitive change in the structure of cellular membranes. It has long been recognised that at least some adaptive processes, most particularly acclimation to temperature and salinity, generally are accompanied by a characteristic change in membrane lipid composition. For many years these changes have been widely accepted, although until recently without any unequivocal supporting evidence, as indicating an important role of lipids in acclimation.

With the development in the late 1960s of sensitive physical-chemical techniques for measuring lipid physical properties, a more logical rationale for examining the role of lipids emerged. It became feasible to obtain fairly precise measurements of those physical properties contributing to the overall fluidity of the membrane and to compare any observed changes with rates of lipid structural alteration (see Cossins, this volume). The fact that lipid alterations measured in these cases tended to offset the environmentally induced changes in membrane fluidity strengthened the hypothesis that a physiologically essential, 'optimum' membrane fluidity is controlled by modifications of structural lipids.

In this review I shall focus attention on the biochemical mechanisms whereby cells can modify the fluidity of their membrane lipids. The term fluidity is used in a general sense to imply that the physical state of the membrane features a variable degree of molecular flexibility and disorder. At the present time no one measurement can adequately depict fluidity in quantitative terms, but certain properties of a membrane contributing to its fluidity can be expressed using such quantitative units as phase transition temperature, polarisation anisotropy, and order parameter (see Cossins, this volume).

Overview of stress-induced membrane lipid alterations

Our present concept of how environmental stress modifies lipid composition derives primarily from a comparison of many cell types that have been cultured for long periods of time and for many cell generations in the presence or in the absence of the stress-producing factor. Considerably fewer organisms have been studied extensively during the initial period of acclimation to a particular environmental stress. Still, a fairly comprehensive picture of cellular lipid responses to stress, especially temperature stress, has emerged. Table 1 summarises the most frequently reported environmentally caused changes in lipid composition. All of these changes can significantly alter membrane fluidity. In the following pages I shall discuss the enzymatic mechanisms whereby these, and one or two less commonly observed alterations, are brought about.

Mechanisms for achieving a homeoviscous response

The fluidity of a biological membrane may be modified by changes in several parts of its component lipids. I shall discuss separately a number of different enzymatic mechanisms all leading to fluidity change.

Changes in the degree of fatty acid unsaturation

The most commonly observed alteration of membrane lipids that is triggered by environmental stress, involves a change in unsaturation of the fatty acids bound to phospholipids, glycolipids, and other structural lipids. This change can come about in a variety of ways, depending upon the organism and the exact nature of the fatty acid. A few examples of stress-induced changes in fatty acid unsaturation are given in Table 2. The table includes only those fatty acids showing the most pronounced change in each case. The extent of change shown in the table is typical of most organisms that have been studied, but there are other cases where much

Table 1. *Common responses of membrane lipid composition to environmental stress*

Type of lipid change	Responsive organisms	Causative factor(s)
Degree of fatty acid unsaturation	Animals, bacteria, higher plants	Temperature, salinity, oxygen tension
Proportions of branched-chain fatty acids	Bacteria	Temperature
Proportions of phospholipid or glycolipid classes	Animals, bacteria, higher plants	Temperature, salinity
Distribution of lipid molecular species	Animals, bacteria	Temperature
Sterol to phospholipid ratio	Animals, plants	Temperature

Table 2. *Examples of stress-induced changes in polar lipid fatty acid unsaturation. Only those fatty acids showing the most pronounced change are listed*

Cell type	Fatty acid	Condition 1	Condition 2	Reference
Cultured mouse LM cells	16:0 18:1	37 °C 20.9 57.8	28 °C 14.2 62.3	Ferguson et al. (1975)
Goldfish brain	18:0 20:4	30 °C 18.3 1.3	5 °C 13.0 4.1	Johnston & Roots (1964)
Tetrahymena cells	16:0 18:2 18:3	39.5 °C 12.6 14.5 24.5	15 °C 8.9 20.2 31.1	Fukushima et al. (1976)
Epicotyls of winter wheat seedlings (Karkov variety)	18:2 18:3	24 °C 27.0 39.4	2 °C 17.8 52.2	de la Roche (1978)
Alfalfa chloroplasts	18:2 18:3	No NaCl 17 67	0.16 M NaCl 24 59	Harzallah-Skhiri et al. (1980)
E. coli (stationary phase) phosphatidylethanolamine	16:1 18:cyclopropane	No NaCl 21.7 15.3	0.3 M NaCl 3.9 32.3	McGarrity & Armstrong (1975)

greater (Fulco, 1972; Sato & Murata, 1980) or much smaller (Gelman & Cronan, 1972; Patterson, Kenrick & Raison, 1978; Mattox & Thompson, 1980) changes have been reported. Many literature reports, including some of those listed in Table 2, furnish data on whole cells or tissues. As I shall discuss later, analyses of whole eukaryotic cells or organs usually provide a very unreliable picture of what alterations have taken place in the various structurally different membranes within a particular cell type. Analyses of whole-cell lipid compositions are therefore of limited usefulness in trying to discover the physiological consequences of stress on membranes.

Fatty acid desaturation in eukaryotic cells. In eukaryotic animal cells, saturated fatty acids are synthesised by a cytoplasmic enzyme system, with the main products being palmitic acid ($C_{16:0}$) and stearic acid ($C_{18:0}$). Further modification of these fatty acids is carried out by a family of microsomal desaturases. The first double bond is inserted by an enzyme specific for a saturated fatty acyl-CoA substrate. One or more additional double bonds may then be added, each by a substrate-specific desaturase. The most common stepwise sequence of desaturation is shown in Fig. 1.

Animals generally create polyunsaturated fatty acids by inserting additional double bonds at positions between the initial unsaturation and the fatty acid's carboxyl group. The exception, linoleic acid, cannot be synthesised by many animals and must be obtained by those species from dietary sources.

Plants differ from animal cells in certain respects, namely, in that most fatty acids are synthesised by the chloroplast (Stumpf, 1980). The initial desaturation step also occurs within the chloroplast, but subsequent double bonds are inserted by microsomal enzymes, as in animal cells, after the monoenes are transferred from the chloroplast. Plant fatty desaturases usually form polyunsaturates by adding additional double bonds towards the end of the chain opposite that containing the carboxyl group (Fig. 2).

There is a strong evidence that most further desaturation of monoenes by both plants and animals involves as substrates fatty acids that are already integrated into phospholipids (Kates & Pugh, 1980). The active sites of these desaturases, and also of those enzymes using fatty acyl-CoA, appear to be buried within the hydrophobic interior of the microsomal membranes. A series of hydrophobic electron carrier proteins furnishes electrons while molecular oxygen serves as a co-substrate for the reaction (Fig. 3).

The extent of phospholipid fatty acid unsaturation increases when the cell is subjected to the stress of low temperatures. Several explanations, some of them trivial, have been proposed to account for this response. A few of the proposals have been tested extensively while others have not. Examples of these findings, often illustrated by experiments carried out with my own favourite model system, the ciliate *Tetrahymena pyriformis*, are described below.

Regulation via the supply of oxygen. Some years ago it was proposed (Harris & James, 1969) that the higher solubility of oxygen in water at low temperature could enhance fatty acid desaturation in plant tissues simply by increasing the availability of the oxygen co-substrate (see Fig. 3) for the desaturase enzyme system. This hypothesis was later tested in my laboratory

Fig. 1. The pathway for fatty acid desaturation in animal tissues.

```
                16:0
               palmitic
              /        \
            C₂          \
           /             \
         18:0           Δ9-16:1
        stearic        palmitoleic
           |
Δ5,8,11,14-20:4
  arachidonic
           ↑
           |
           |
         Δ9-18:1
          oleic
           |
Δ8,11,14-20:3
           ↑
           |
         Δ9,12-18:2
          linoleic
           |
         C₂
           |
Δ6,9,12-18:3 ←———┘
  γ-linolenic
```

Fig. 2. The pathway for fatty acid desaturation in plant tissues.

```
                16:0
               palmitic
              /        \
            C₂          \
           /             \
         18:0          Δ7-16:1
        stearic
           |              |
         Δ9-18:1       Δ7,10-16:2
          oleic
           |              |
         Δ9,12-18:2    Δ7,10,13-16:3
          linoleic
           |
         Δ9,12,15-18:3
          α-linolenic
```

Fig. 3. The microsomal transport of electrons for fatty acid desaturation. CSF, cyanide-sensitive factor; fp$_1$, NADH–cytochrome b_5 reductase; fp$_2$, NADPH-specific flavoprotein.

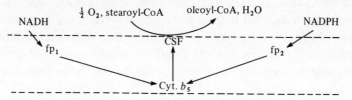

by exposing *Tetrahymena pyriformis* cells to a wide range of oxygen concentrations either isothermally or at different temperatures (Skriver & Thompson, 1976). In all cases the resulting fatty acid composition was very characteristic of the temperature but not the oxygen tension, indicating that even at quite low cellular levels of oxygen, its availability is not a limiting factor in the rate of fatty acid desaturation.

On the other hand, recent work by Rebeille, Bligny & Douce (1980) established that the level of fatty acid unsaturation in sycamore cells grown in culture was highly dependent upon the concentration of dissolved oxygen. Particularly noteworthy was the dramatic reversal from a high oleic acid ($C_{18:1}$) and low linoleic acid ($C_{18:2}$) content at low oxygen tension to a low $C_{18:1}$ and high $C_{18:2}$ content at high oxygen tension. Temperature change *per se* had little effect on fatty acid unsaturation in these cells. Thus the sensitivity to oxygen appears to differ greatly from species to species.

Regulation via the supply of saturated fatty acids. It is known from work with many animals, including *Tetrahymena* (Martin *et al.*, 1976), that chilling sharply reduces the rate of fatty acid synthesis. The rate of desaturation is also inhibited, in absolute terms. If the inhibition of synthesis were greater than that of desaturation, then one might logically expect the degree of fatty acid unsaturation to increase. However, it can easily be shown that the elevated level of unsaturated fatty acids at low temperature is more than a simple result of supply and demand. For example, *Tetrahymena* cells can incorporate relatively large amounts of exogenous saturated fatty acids added either just before or during chilling. Even with this abnormally plentiful supply of saturated fatty acids being drawn into phospholipids, increase in unsaturation at low temperature proceeds in the usual way (G. A. Thompson, Jr, unpublished).

Regulation by the induced synthesis of fatty acid desaturases. As I shall describe later, certain prokaryotes, such as *Bacillus megaterium*, lack even traces of fatty acid desaturase activity at higher temperatures, but rapidly synthesise the enzyme when chilled (Fulco, 1972). Efforts to find a similar response in eukaryotic cells have been partly successful. In chilled *Tetrahymena*, palmitoyl-CoA desaturase activity has been shown to increase by a process that is sensitive to inhibitors of protein synthesis (Nozawa & Kasai, 1978). No definitive evidence is available to suggest that the synthesis of other *Tetrahymena* desaturases or any desaturases of other eukaryotes is inducible by low temperature. But there is good support (Oshino & Sato, 1972) for a rapid synthesis of fatty acid desaturases in the liver of fasted rats induced by re-feeding carbohydrates. Thus the ability of cells to increase their complement of fatty acid desaturating enzymes when needed may be an important factor in acclimation to stress.

Regulation of desaturases by membrane fluidity changes. It has been proposed that the relative activity of the fatty acid desaturases may depend upon the fluidity of their membrane environment (Kasai *et al.*, 1976). Evidence supporting this concept includes the observation that a variety of treatments which alter membrane fluidity (e.g. feeding unsaturated fatty acids (Kasai *et al.*, 1976) or exposing cells to the fluidising effects of general anaesthetics (Nandini-Kishore, Kitajima & Thompson, 1977)) could effectively modulate the activity of the fatty acid desaturase in *Tetrahymena* and could, in some cases, overcome the desaturase-activating effects of low temperature. Physical constraints imposed upon one or more of the microsomal desaturases or members of their associated electron transport system by decreased fluidity might be envisioned as promoting enhanced desaturase activity, either by triggering a protein conformational change or by reorienting the enzyme's active site with respect to its surrounding phospholipid substrate. There is evidence for a fluidity-induced physical movement of some integral proteins perpendicular to the plane of the membrane (Wunderlich *et al.*, 1975), but a direct involvement of this movement with altered desaturase activity has not been proved.

Fatty acid desaturation in prokaryotic cells. As in eukaryotes, many bacteria and blue-green algae respond to low temperature by increasing the proportion of unsaturated fatty acids in their membrane lipids (Thompson, 1980*a*). However, the mechanism for achieving this change is, in some species, quite different from that observed in eukaryotic organisms. Anaerobic bacteria, such as *Escherichia coli*, control the rate of unsaturated fatty acid synthesis by regulating an enzyme which also participates in the formation of saturated fatty acids. At low temperatures the enzyme β-hydroxydecanoyl thioester dehydrase converts more of its substrate into the *cis*-3 isomer of decenoate, which is then further elongated to yield an unsaturated fatty acid (Fig. 4). The precise reason for this increased production of the *cis*-3 isomer in preference to the *trans*-2 isomer at low temperature is not clearly established, but it may result from a selective utilisation of a *cis*-3-decenoate product by one of the enzymes participating in a subsequent elongation step (Garwin, Klages & Cronan, 1980).

Aerobic bacteria form mono-unsaturated fatty acids by a totally different mechanism. As in eukaryotic cells, the double bond is introduced into pre-formed saturated fatty acids with the participation of oxygen. In *Bacillus megaterium*, extensive studies by Fulco and co-workers (Fujii & Fulco, 1977) have proved that exposure to low temperature induces the synthesis of a $\Delta 5$-desaturase. Control is exerted through a protein modulator promoting desaturase synthesis. The modulator is unstable at high temperature and is degraded if the cells are warmed to 35 °C.

Changes in the proportion of branched-chain fatty acids

The fluidising effect of the *cis* double bond in fatty acids is simulated to some extent by methyl-branched fatty acids. Both iso-branched and anteiso-branched fatty acids occur widely as components of membrane lipids, particularly in bacteria (Kaneda, 1977). The anteiso fatty acids melt about 25–35 deg C lower than fatty acids of the normal series and sometimes seem to be used for the same adaptive purpose as are unsaturated fatty acids. Some

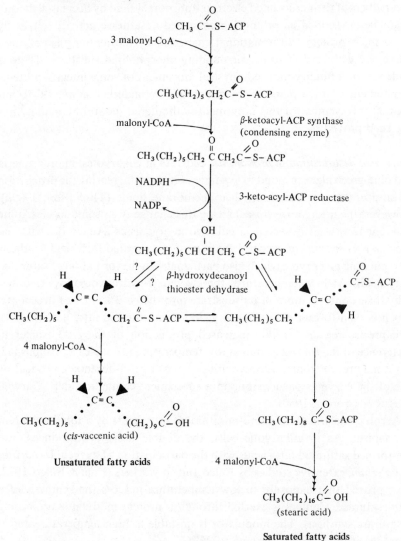

Fig. 4. The biosynthesis of saturated and unsaturated fatty acids in *Escherichia coli*. (Modified from Thompson, 1980a.)

species of the genus *Bacillus* have been shown to contain much higher levels of anteiso- and iso-branched chain fatty acids when grown at low temperatures than at higher temperatures.

The biosynthesis of branched-chain fatty acids differs from that of normal fatty acids in that the initial condensation step involves a branched-chain-CoA initiator rather than acetyl-CoA. *Bacillus* species appear to have two fatty acid synthetase systems, one producing straight-chain and the other branched-chain products. The mechanism whereby the activities of these two systems are regulated is unknown.

Changes in fatty acid chain length

Lipids from cells grown under environmental stress often have an average fatty acid chain length somewhat different from that of non-stressed cells. These differences are usually less striking than are those involving fatty acid unsaturation, and chain length has not generally been considered as a factor of prime importance.

However, in some cases the modification of fatty acid chain length outweighs all other changes. Thus the gram-negative psychrophilic bacterium *Micrococcus cryophilus*, whose lipids contain only 18:1 and 16:1 fatty acids, responds to a lowering of the growth temperature from 20 °C to 0 °C by shifting its C_{18}/C_{16} ratio from 3.7 to 0.8 (Russell & Sandercock, 1980). It was postulated that the processes of fatty acid elongation and retroconversion are regulated by membrane fluidity in such a way as to favour the shorter chain products at low temperature.

Changes in the positioning of fatty acids in membrane lipids

Just as important as the types of fatty acid present in membranes is their precise location in specific lipids. The recent increasing use of sophisticated analytical techniques has made it evident that fatty acid placement in individual lipid molecules is anything but random. Enzymes responsible for this non-random fatty acid positioning seem to be affected by environmental stress.

It may be appropriate at this point to emphasise that functionally different membranes within a given cell usually have quite distinctive lipid compositions. Studies correlating lipid composition with membrane physical properties should utilise as nearly homogeneous membrane preparations as possible. Lipid changes observed in extracts of whole cells or crude fractions thereof may result from undecipherable alterations within several of the membrane types present or simply from changes in the relative proportions of the different membranes, each of which, taken alone, would exhibit no change in lipid composition.

Knowing in quantitative terms the exact phospholipid or glycolipid molecular species occurring in a homogeneous membrane is the ultimate goal which has only recently become possible. However, in the few cases where this information is already available, we cannot yet interpret its significance in terms of physical properties because of the complexity of the mixtures. But it is plainly at this level that such a correlation ultimately needs to be made.

Probably the most sensitive technique for phospholipid molecular species analysis is combined gas chromatography and mass spectrometry (GCMS) (Satouchi & Saito, 1979; Kuksis & Myher, 1980). Each phospholipid class separated by thin-layer chromatography is hydrolysed to diglycerides (or ceramides, in the case of sphingolipids) with phospholipase C, and the diglycerides are then converted to *tert*-butyldimethylsilyl ether derivatives. These derivatives can be identified and quantified by GCMS with rapidly improving accuracy. Complementary phospholipase A studies through which the specific fatty acids located at the sn-1 position of the phospholipid glycerol moiety can be differentiated from those at the sn-2 position complete the molecular species analysis.

We have shown the reorganisation of phospholipid molecular species to be particularly important during the initial stages of acclimation to low-temperature stress by *Tetrahymena* microsomes (Dickens & Thompson, 1982). When *Tetrahymena* growing at 39 °C were rapidly chilled to 15 °C, growth resumed only after an acclimation period of 15 hours. However, certain fluidity-related physical properties of microsomal lipids, as estimated by fluorescence polarisation measurements, changed within 1 hour from the 39 °C pattern to the 15 °C pattern (Dickens & Thompson, 1981). During this 1-hour period the microsomal phospholipid distribution was unchanged, and the fatty acid composition of each phospholipid class experienced only slight changes. GCMS analysis of each phospholipid class revealed a previously unsuspected reorganisation of fatty acids on the basis of chain length (as well as unsaturation) so as to enhance certain combinations. In some cases the changes in carbon number (combined chain length of the two component fatty acids) were considerably greater than the change in unsaturation during this initial phase of acclimation (Table 3). Further analysis of the derivatised diglycerides by GCMS confirmed that the retailoring process led specifically to certain preferred molecular species. The complexity of the natural mixture (over 60 quantitatively significant species) has so far prevented a clear interpretation of how the observed changes affect membrane fluidity in this system, but the findings do suggest that retailoring plays an important role.

Table 3. *Comparative effects of brief chilling on the fatty acid composition and the intramolecular pairing of fatty acids in* Tetrahymena *microsomal phosphatidylethanolamine*

Fatty acids			Combined fatty acid chain length		
	Temperature			Temperature	
Fatty acid	39 °C-grown	39 °C → 15 °C (1 hour)	Carbon no.	39 °C-grown	39 °C → 15 °C (1 hour)
12:0	1.2±0.2	0.7±0.2	26	1.1±0.1	1.7±0.5
14:0	17.8±0.2	15.3±1.3	27	1.5±0.3	1.2±0.4
15:0 (iso)	6.1±0.1	5.5±0.2	28	5.3±0.6	3.7±0.4
15:0	3.2±0.1	3.3±0.1	29	4.1±0.6	3.4±1.4
16:0	14.4±0.1	12.0±0.4	30	15.2±3.8	19.9±0.6
16:1	20.3±0.1	21.6±0.6	31	11.1±0.6	16.3±0.1
16:2 + 17:1	7.7±0.3	7.5±0.1	32	23.3±1.9	26.5±2.5
18:0	2.7±0.1	2.7±0.2	33	12.7±0.4	7.3±1.3
18:1	4.9±0.4	4.5±0.6	34	20.6±2.8	11.5±1.4
18:2	8.8±0.2	9.7±1.0	36	3.2±2.9	7.7±3.3
18:3	10.1±0.2	13.0±0.1			

B. F. Dickens & G. A. Thompson (unpublished).
Values represent % composition.

Changes in the relative proportions of complex lipid classes

Apart from the striking changes in fatty acid unsaturation, alterations in the distribution of the polar head groups of lipids are perhaps the most frequently reported membrane response to environmental stress. Typical of the observed differences are the values shown in Table 4. On the basis of the few physical-chemical studies that have been done, differences of this magnitude might be expected to have a perceptible effect on membrane fluidity. Synthetic phosphatidylcholine can have a phase transition temperature 25 deg C lower than phosphatidylethanolamine containing the same fatty acids (Oldfield & Chapman, 1972), and digalactosyldiglycerides are considerably more fluid than the equivalent molecular species of monogalactosyldiglycerides (Bishop *et al.*, 1980). The regulatory mechanisms responsible for the stress-induced changes in polar head groups are not understood. In at least one case, *Tetrahymena*, the change in polar head groups (Table 4) does not take place during the first hours of low-temperature acclimation, when changes in the phospholipid hydrocarbon acyl chains are most evident. Only after cell growth resumes at the lower temperature do the alterations in polar head groups commence. This might indicate a secondary role for these latter modifications, perhaps correcting localised imbalances in membrane fluidity created by earlier alterations (Fig. 5) (Israelachvili, Marčelja & Horn, 1980).

Unfortunately, it is not yet possible to set down a general rule whereby stress-induced changes in polar head groups can be predicted. According to the lipid physical properties mentioned above, the decreases in phosphatidylethanolamine and monogalactosyldiglyceride should indeed contribute towards fluidisation of the membrane at low temperature, but any such effect of polar constituents would be strongly influenced by even small fatty acid changes that might accompany them.

Likewise, it is impossible to generalise on the regulatory mechanisms underlying these changes. Even the more obvious modifications, such as the apparent conversion of monogalactosyldiglyceride to digalactosyldiglyceride in chilled *Dunaliella* (Table 4), may not be as straightforward as they seem. In that particular study, the fatty acid compositions of the two glycolipid classes were so dissimilar that a simple glycosylation of one class to produce the other has to be questioned.

Besides analytical data on homogeneous membrane preparations, such as those exemplified in Tables 2 and 4, the literature contains numerous reports describing sizable compositional changes in whole cells and tissues. In the absence of more refined data, it is tempting to speculate as to the significance of these changes in polar head groups. In a few cases, patterns of behaviour seem to follow a common plan. Plant tissues, for example, often experience an increase in phospholipids, especially phosphatidylethanolamine, when

Table 4. *Effects of environmental stress on lipid polar head group distribution. Only the more pronounced changes are shown in each case*

System	Stress	Principal changes (% of total phospholipid or glycolipid)			Reference
Goldfish red muscle mitochondria	Temperature	Phosphatidylcholine Phosphatidylethanolamine	30 °C 43.7 39.3	5 °C 56.9 24.9	Addink (1980)
Tetrahymena pyriformis microsomes	Temperature	2-Aminoethylphosphonolipid Phosphatidylethanolamine	39 °C 14.7 43.9	15 °C 22.9 34.1	Fukushima *et al.* (1976)
Dunaliella salina chloroplasts	Temperature	Monogalactosyldiglyceride Digalactosyldiglyceride	30 °C 66 19	12 °C 58 28	Lynch & Thompson (1982)
Staphylococcus aureus cells	Salinity	Cardiolipin Phosphatidylglycerol Lysylphosphatidylglycerol	10% NaCl 50 39 5	0.05% NaCl 10 65 12	Kanemasa *et al.* (1972)

exposed to chilling or freezing temperatures (Willemot, 1979). A probable explanation for this response is suggested by our recent observations of chilled *Dunaliella salina* cells (Lynch & Thompson, 1982). Morphometric studies of cells grown at 30 °C and 12 °C revealed an almost three-fold increase in the relative amount of endoplasmic reticulum. The proliferation of this membrane, which is rich in phosphatidylethanolamine, caused a significant increase in the ratio of phosphatidylethanolamine to other phospholipids in total cell extracts, although the individual cell organelles each maintained a virtually constant composition of polar head groups.

Changes in the cellular content of sterols

As described by Cossins in this volume, the addition of sterols to a phospholipid mixture reduces its sensitivity to temperature-induced fluidity change. Cholesterol has been postulated as serving an important function in the physiological regulation of membrane physical properties (Melchior, Scavitto & Steim, 1980).

There have been few reports of changes in membrane sterol content consistent with it playing a significant role in homeoviscous adaptation. Little change was found in the cholesterol content of synaptosomal membranes of goldfish adapted to different water temperatures (Cossins, 1977), and

Fig. 5. The effect of polar head group size on the three-dimensional shape and stability of membrane lipids. (Modified from Israelachvili *et al.*, 1980.)

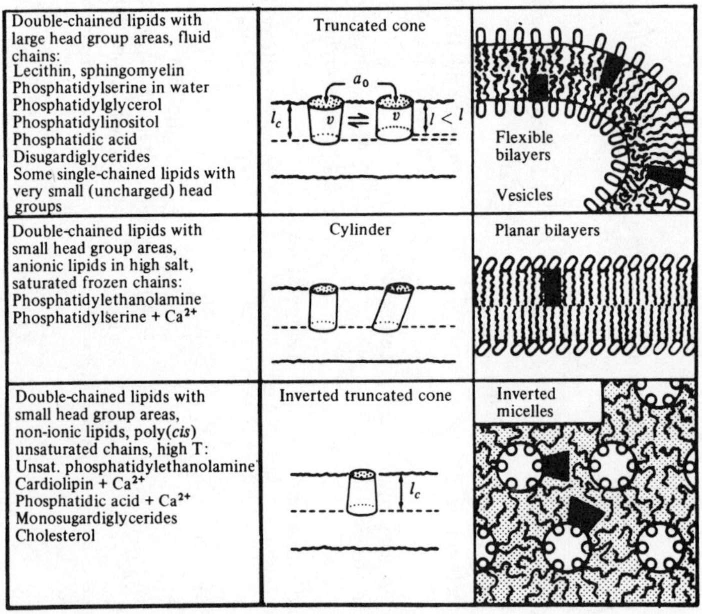

Double-chained lipids with large head group areas, fluid chains: Lecithin, sphingomyelin Phosphatidylserine in water Phosphatidylglycerol Phosphatidylinositol Phosphatidic acid Disugardiglycerides Some single-chained lipids with very small (uncharged) head groups	Truncated cone	Flexible bilayers Vesicles
Double-chained lipids with small head group areas, anionic lipids in high salt, saturated frozen chains: Phosphatidylethanolamine Phosphatidylserine + Ca^{2+}	Cylinder	Planar bilayers
Double-chained lipids with small head group areas, non-ionic lipids, poly(*cis*) unsaturated chains, high T: Unsat. phosphatidylethanolamine Cardiolipin + Ca^{2+} Phosphatidic acid + Ca^{2+} Monosugardiglycerides Cholesterol	Inverted truncated cone	Inverted micelles

Tetrahymena microsomes maintained a constant ratio of tetrahymanol (a sterol analogue) to phospholipid during low-temperature acclimation (Fukushima *et al.*, 1976). On the other hand, the molar ratio of cholesterol to phospholipid in carp liver mitochondria decreased from 0.12 at an environmental temperature of 32 °C to 0.09 at 10 °C (Wodtke, 1978). And it was recently reported that a preparation of crude membranes from Chinese hamster ovary cells exhibited significantly higher cholesterol to phospholipid ratios as the temperature at which the cells had been incubated was increased from 32 °C to 41 °C (Anderson *et al.*, 1981). It will be interesting to find out whether the observed increase in the relative cholesterol content in this latter report reflects an actual change in the cholesterol to phospholipid ratio in certain membrane types or a change in the relative proportions of cholesterol-rich and cholesterol-poor organelles.

Temporal relationships among cellular membranes during adaptation

A cell may possess one or more of the above-mentioned mechanisms for membrane lipid modification and still not survive even a moderate stress if that stress is imposed suddenly. Because eukaryotic cells are highly compartmentalised, most lipid modifications are brought about in one region of the cell, usually the endoplasmic reticulum. Time is required to (1) make the necessary lipid changes, and (2) disseminate these altered lipids to other cellular membranes. If the time needed to complete the adaptive changes is slow compared with the time over which the stress is applied, the cell may be unable to overcome the physiological problems caused by the stress. Fig. 6 portrays two scenarios commonly encountered in nature. Slow cooling of a plant or poikilothermic animal will allow time for internal lipid changes to compensate for any temperature-induced lowering of membrane fluidity. On the other hand, cooling that is rapid with respect to the response time of the organism can lead to an inactivation of membrane functions so extensive and prolonged that fluidisation of the lipids cannot be completed.

Little detailed information is available on the dynamics of the acclimation process. We have made a special effort to examine this point using *Tetrahymena* (Thompson, 1980*b*). It became apparent in early studies that a rapid desaturation of phospholipid fatty acids occurred in the microsomes, leading to a significant fluidisation of that membrane within 1 hour (Fig. 7). But a relatively slow dissemination of the modified lipids into the cell surface (pellicle) membranes delayed any sizable increase in their fluidity. The ciliary membrane, which is also associated with the cell surface, is even slower to reflect the trend of increased fatty acid unsaturation after chilling (Ramesha & Thompson, 1982). These delays with respect to intracellular lipid movement may contribute significantly to the fact that growth of chilled *Tetrahymena* cultures resumes only after 12 to 15 hours of incubation at the low temperature.

Observations such as these, suggest that the rate-limiting step in low-temperature acclimation in at least some cell types may be the transport of lipids from the intracellular site at which they are structurally modified to other membrane destinations. It is unlikely that this movement is achieved by an actual flow of intact membrane elements (Morré & Ovtracht, 1977), since net growth seldom occurs under stress. More probably, lipids move via an exchange process, such as that known to be mediated by phospholipid exchange proteins (Zilversmit & Hughes, 1976; Mazliak & Kader, 1980). Since the specificity of phospholipid exchange proteins appears to be influenced by the fluidity of the donor and acceptor membrane (Helmkamp, 1980), the possibility of an exchange directed predominantly into certain organelles has to be considered.

Conclusions

Considering the variety of independent biochemical pathways which can be utilised to alter membrane fluidity in environmentally stressed cells,

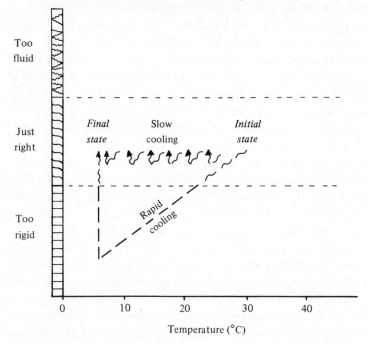

Fig. 6. Representation of an organism's ability to maintain its membrane fluidity within an optimum range during a slow temperature reduction but not a fast one. (Modified from Thompson, 1979.)

it is hardly surprising that so much mystery still surrounds the phenomenon of homeoviscous adaptation. Some organisms rely heavily on one or more mechanisms which are totally unlike those employed by a different species. It is still not straightforward to recognise whether a physiologically significant alteration has in fact been made, because sizable changes in membrane fluidity have now been found to result from lipid modifications so subtle as to escape notice in a routine lipid analysis.

I believe that recent experience in the field has provided useful guidelines that we must heed in order to assure future progress. We have seen how different membranes, even within the same cell, can respond to stress in different ways and at different rates. This awareness will give us greater incentive to focus attention on detailed changes occurring within each homogeneous membrane population. The availability of improved cell fractionation schemes and a wise choice of model systems should render this feasible.

We have also come to realise that predicting the effect of a particular change in lipid composition on membrane fluidity is not always easy. More and more

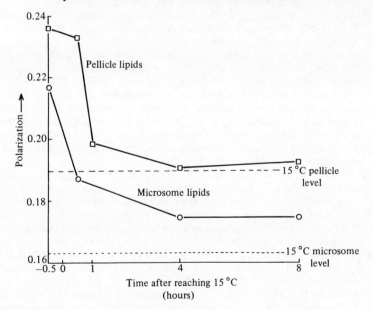

Fig. 7. Time-course of changes in diphenylhexatriene polarisation in membrane lipids of 39.5 °C-acclimated *Tetrahymena* cells following a shift to 15 °C over a 30-minute period. At the times indicated, cells were fractionated and total lipids from purified pellicles and microsomes used for polarisation measurements. Decreasing polarisation signifies increasing fluidity. The dashed lines show polarisation values found in the two fractions from cells fully acclimated to 15 °C. For details see Martin & Thompson (1978).

often will it be necessary actually to demonstrate a finite effect, or the lack of one, using at least one but preferably two or more independent physical techniques capable of sensing fluidity-related parameters. I suspect we may discover through this approach that several inconspicuous compositional changes sometimes combine to produce a synergistic effect of real physiological importance.

There was a time, not so long ago, when many of us held stubbornly to the idea of membrane adaptation to environmental stress being accomplished in all creatures by some common mechanism. While I still believe that changes in membrane lipids have definite survival value, I have finally abandoned my dogmatic view of how they must be achieved. For my future research, that change in perspective may work wonders.

Experiments conducted in the author's laboratory were supported in part by the Robert A. Welch Foundation, the National Science Foundation, and the National Institutes of Health.

References

Addink, A. D. F. (1980). Activity of membrane-bound enzymes of the respiratory chain during adaptation of fish to temperature changes. In *Membrane Fluidity. Biophysical Techniques and Cellular Regulation*, ed. M. Kates & A. Kuksis, pp. 99–104. Clifton, New Jersey: Humana Press.

Anderson, R. L., Minton, K. W., Li, G. C. & Hahn, G. M. (1981). Temperature-induced homeoviscous adaptation of Chinese hamster ovary cells. *Biochimica et Biophysica Acta*, **641**, 334–48.

Bishop, D. G., Kenrick, J. R., Bayston, J. H., Macpherson, A. S. & Johns, S. R. (1980). Monolayer properties of chloroplast lipids. *Biochimica et Biophysica Acta*, **602**, 248–59.

Cossins, A. R. (1977). Adaptation of biological membranes to temperature. The effect of temperature acclimation of goldfish upon the viscosity of synaptosomal membranes. *Biochimica et Biophysica Acta*, **470**, 395–411.

de la Roche, I. A. (1978). Development of freezing tolerance in wheat without changes in lipid unsaturation. *Acta Horticulturae*, **81**, 85–9.

Dickens, B. F. & Thompson, G. A., Jr (1981). Rapid membrane response during low temperature acclimation. Correlation of early changes in the physical properties and lipid composition of microsomal membranes. *Biochimica et Biophysica Acta*, **644**, 211–18.

Dickens, B. F. & Thompson, G. A. Jr (1982). Phospholipid molecular species alterations in microsomal membranes as an initial key step during cellular acclimatisation to low temperature. *Biochemistry*, **21**, 3604–11.

Ferguson, K. A., Glaser, M., Bayer, W. H. & Vagelos, P. R. (1975). Alteration of fatty acid composition of LM cells by lipid supplementation and temperature. *Biochemistry*, **14**, 146–51.

Fujii, D. K. & Fulco, A. J. (1977). Biosynthesis of unsaturated fatty acids by bacilli. Hyperinduction and modulation of desaturase synthesis. *Journal of Biological Chemistry*, **252**, 3660–70.

Fukushima, H., Martin, C. E., Iida, H., Kitajima, Y., Thompson, G. A., Jr & Nozawa, Y. (1976). Changes in membrane lipid composition during temperature adaptation by a thermotolerant strain of *Tetrahymena pyriformis*. *Biochimica et Biophysica Acta*, **431**, 165–79.

Fulco, A. J. (1972). The biosynthesis of unsaturated fatty acids by bacilli. IV. Temperature-mediated control mechanisms. *Journal of Biological Chemistry*, **247**, 3511–19.

Garwin, J. L., Klages, A. L. & Cronan, J. E., Jr (1980). β-Ketoacyl-acyl carrier protein synthase II of *Escherichia coli*. Evidence for function in the thermal regulation of fatty acid synthesis. *Journal of Biological Chemistry*, **255**, 3263–5.

Gelman, E. P. & Cronan, J. E., Jr (1972). Mutant of *Escherichia coli* deficient in the synthesis of *cis*-vaccenic acid. *Journal of Bacteriology*, **112**, 381–7.

Harris, P. & James, A. T. (1969). The effects of low temperatures on fatty acid biosynthesis in plants. *Biochemical Journal*, **112**, 325–30.

Harzallah-Skhiri, F., Guillot-Salomon, T. & Signol, M. (1980). Lipid changes in plastids isolated from alfalfa seedlings grown under salt stress. In *Biogenesis and Function of Plant Lipids*, ed. P. Mazliak, P. Benveniste, C. Costes & R. Douce, pp. 99–102. Amsterdam: Elsevier.

Helmkamp, G. M., Jr (1980). Effects of phospholipid fatty acid composition and membrane fluidity on the activity of bovine brain phospholipid exchange protein. *Biochemistry*, **19**, 2050–6.

Israelachvili, J. N., Marčelja, S. & Horn, R. G. (1980). Physical principles of membrane organisation. *Quarterly Reviews of Biophysics*, **13**, 121–200.

Johnston, P. V. & Roots, B. I. (1964). Brain lipid fatty acids and temperature acclimation. *Comparative Biochemistry and Physiology*, **11**, 303–9.

Kaneda, T. (1977). Fatty acids of the genus *Bacillus*: an example of branched chain preference. *Bacteriological Reviews*, **41**, 391–418.

Kanemasa, Y., Yoshioka, T. & Hayashi, H. (1972). Alteration of the phospholipid composition of *Staphylococcus aureus* cultured in medium containing NaCl. *Biochimica et Biophysica Acta*, **280**, 444–50.

Kasai, R., Kitajima, Y., Martin, C. E., Nozawa, Y., Skriver, L. & Thompson, G. A., Jr (1976). Molecular control of membrane properties during temperature acclimation. Membrane fluidity regulation of fatty acid desaturase action? *Biochemistry*, **15**, 5228–33.

Kates, M. & Pugh, E. (1980). Role of phospholipid desaturases in control of membrane fluidity. In *Membrane Fluidity. Biological Techniques and Cellular Regulation*, ed. M. Kates & A. Kuksis, pp. 153–70. Clifton, New Jersey: Humana Press.

Kuksis, A. & Myher, J. J. (1980). New approaches to lipid analysis of lipoproteins and cell membranes. In *Membrane Fluidity. Biophysical Techniques and Cellular Regulation*, ed. M. Kates & A. Kuksis, pp. 3–26. Clifton, New Jersey: Humana Press.

Lands, W. E. M. (1980). Dialogue between membranes and their lipid metabolizing enzymes. *Transactions of the Biochemical Society*, **8**, 25–7.

Lynch, D. V. & Thompson, G. A., Jr (1982). Low-temperature-induced alterations in the chloroplast and microsomal membranes of *Dunaliella salina*. *Plant Physiology*, **69**, 1369–75.

McGarrity, J. T. & Armstrong, J. B. (1975). The effect of salt on phospholipid fatty acid composition in *Escherichia coli* K-12. *Biochimica et Biophysica Acta*, **398**, 258–64.

Martin, C. E., Hiramitsu, K., Kitajima, Y., Nozawa, Y., Skriver, L. & Thompson, G. A., Jr (1976). Molecular control of membrane properties during temperature acclimation. Fatty acid desaturase regulation of membrane fluidity in acclimating *Tetrahymena* cells. *Biochemistry*, **15**, 5218–27.

Martin, C. E. & Thompson, G. A., Jr (1978). Use of fluorescence polarization to monitor intracellular membrane changes during temperature acclimation. Correlation with lipid compositional and ultrastructural changes. *Biochemistry*, **17**, 3581–6.

Mattox, S. M. & Thompson, G. A., Jr (1980). The effects of high concentrations of sodium or calcium ions on the lipid composition and properties of *Tetrahymena* membranes. *Biochimica et Biophysica Acta*, **599**, 24–31.

Mazliak, P. & Kader, J. C. (1980). Phospholipid exchange systems. In *The Biochemistry of Plants*, vol. 4, *Lipids, Structure and Function*, ed. P. K. Stumpf, pp. 283–300. New York & London: Academic Press.

Melchior, D. L., Scavitto, F. J. & Steim, J. J. (1980). Dilatometry of dipalmitoyllecithin–cholesterol bilayers. *Biochemistry*, **19**, 4828–34.

Morré, D. J. & Ovtracht, L. (1977). Dynamics of the Golgi apparatus: membrane differentiation and membrane flow. *International Review of Cytology* (Supplement 5), 61–188.

Nandini-Kishore, S. G., Kitajima, Y. & Thompson, G. A., Jr (1977). Membrane fluidizing effects of the general anesthetic methoxyflurane elicit an acclimation response in *Tetrahymena*. *Biochimica et Biophysica Acta*, **471**, 157–61.

Nozawa, Y. & Kasai, R. (1978). Mechanism of thermal adaptation of membrane lipids in *Tetrahymena pyriformis* NT-1. Possible evidence for temperature-mediated induction of palmitoyl-CoA desaturase. *Biochimica et Biophysica Acta*, **529**, 54–66.

Oldfield, E. & Chapman, D. (1972). Dynamics of lipids in membranes: heterogeneity and the role of cholesterol. *FEBS Letters*, **23**, 285–97.

Oshino, N. & Sato, R. (1972). The dietary control of the microsomal stearyl CoA desaturation enzyme system in rat liver. *Archives of Biochemistry and Biophysics*, **149**, 369–77.

Patterson, B. D., Kenrick, J. R. & Raison, J. K. (1978). Lipids of chill-sensitive and resistant *Passiflora* species: fatty acid composition and temperature dependence of spin label motion. *Phytochemistry*, **17**, 1089–92.

Ramesha, C. S. & Thompson, G. A., Jr (1982). Changes in the lipid composition and physical properties of *Tetrahymena* ciliary membranes following low temperature acclimation. *Biochemistry*, **21**, 3612–17.

Rebeille, F., Bligny, R. & Douce, R. (1980). Oxygen and temperature effects of the fatty acid composition of sycamore cells (*Acer pseudoplatanus* L.). In *Biogenesis and Function of Plant Lipids*, ed. P. Mazliak, P. Benveniste, C. Costes & R. Douce, pp. 203–6. Amsterdam: Elsevier.

Russell, N. J. & Sandercock, S. P. (1980). The regulation of bacterial membrane fluidity by modification of phospholipid fatty acyl chain length. In *Membrane Fluidity. Biophysical Techniques and Cellular Regulation*, ed. M. Kates & A. Kuksis, pp. 181–90. Clifton, New Jersey: Humana Press.

Sato, N. & Murata, N. (1980). Desaturation of fatty acids in lipids in response to the growth temperature in the blue-green alga, *Anabaena variabilis*. In *Biogenesis and Function of Plant Lipids*, ed. P. Mazliak, P. Benveniste, C. Costes & R. Douce, pp. 207–10. Amsterdam: Elsevier.

Satouchi, K. & Saito, K. (1979). Use of t-butyldimethylchlorosilane/imidazole reagent for identification of molecular species of phospholipids by gas–liquid chromatography/mass spectrometry. *Biomedical Mass Spectrometry*, **6**, 396–402.

Skriver, L. & Thompson, G. A., Jr (1976). Environmental effects on *Tetrahymena* membranes. Temperature-induced changes in membrane fatty acid unsaturation are independent of the molecular oxygen concentration. *Biochimica et Biophysica Acta*, **431**, 180–8.

Stumpf, P. K. (1980). Biosynthesis of saturated and unsaturated fatty acids. In *The Biochemistry of Plants*, vol. 4, *Lipids, Structure and Function*, ed. P. K. Stumpf, pp. 177–204. New York & London: Academic Press.

Thompson, G. A., Jr (1979). Molecular control of membrane fluidity. In *Low Temperature Stress in Crop Plants. The Role of the Membrane*, ed. J. M. Lyons, D. Graham & J. K. Raison, pp. 347–63. New York & London: Academic Press.

Thompson, G. A., Jr (1980a). *The Regulation of Membrane Lipid Metabolism*. Boca Raton: CRC Press.

Thompson, G. A., Jr (1980b). Regulation of membrane fluidity during temperature acclimation by *Tetrahymena pyriformis*. In *Membrane Fluidity. Biophysical Techniques and Cellular Regulation*, ed. M. Kates & A. Kuksis, pp. 381–97. Clifton, New Jersey: Humana Press.

Willemot, C. (1979). Chemical modification of lipids during frost hardening of herbaceous species. In *Low Temperature Stress in Crop Plants. The Role of the Membrane*, ed. J. M. Lyons, D. Graham & J. K. Raison, pp. 411–30. New York & London: Academic Press.

Wodtke, E. (1978). Lipid adaptation in liver mitochondrial membranes of carp acclimated to different environmental temperatures. Phospholipid composition, fatty acid pattern, and cholesterol content. *Biochimica et Biophysica Acta*, **529**, 280–91.

Wunderlich, F., Ronai, A., Speth, V., Seelig, J. & Blume, A. (1975). Thermotropic lipid clustering in *Tetrahymena* membranes. *Biochemistry*, **14**, 3730–5.

Zilversmit, D. B. & Hughes, M. E. (1976). Phospholipid exchange between membranes. *Methods in Membrane Biology*, **7**, 211–59.

ELSE K. HOFFMANN
Volume regulation by animal cells

The volume of animal cells is precisely regulated. This regulation is a fundamental cellular function and malfunctions can lead to cell swelling and lysis. Four physical principles are involved in the determination of cell volume: (1) water is in thermodynamic equilibrium in the system, i.e. in *osmotic equilibrium*; (2) *hydrostatic pressure differences* between the inside and the outside of the cell are *negligible*, since membranes of animal cells are easily distensible and cannot maintain pressure gradients; (3) *electroneutrality* is maintained in the outer as well as in the inner compartment; (4) the permeable ions approach a distribution where the products of activities of anions and cations on both sides of the membrane are equal, i.e. *Gibbs–Donnan distribution*.

Water can move readily through the plasma membrane of most cells whereas the cell matrix contains 'fixed anions' of proteins and nucleic acids that are too large to pass through the membrane. Besides these, the system contains small ions like Na^+, K^+, Cl^- and HCO_3^- to which the membrane is selectively permeable. One might predict that such a system should in isosmotic medium undergo colloid osmotic swelling and ultimately cytolysis (see Hoffman, 1958). This is prevented in living cells by active adjustment of the intracellular content of the diffusible cations Na^+ and K^+, by the sodium/potassium pump. Cl^- is, in many cells, also out of electrochemical equilibrium, due to the activity of an Na^+/Cl^- co-transport mechanism.

Moreover a number of vertebrate cell types can regulate their volume in anisotonic media. These cells initially shrink or swell osmotically, but with continued incubation the volume then reverts towards the original value. This adaptive response was first characterised in frog skin epithelium (MacRobbie & Ussing, 1961), nucleated erythrocytes (Fugelli, 1967; Kregenow, 1971a, b) and in mammalian cell lines (Roti Roti & Rothstein, 1973; Hendil & Hoffmann, 1974). In all these cell types corrective changes in cell size resulted from shifts in cell water brought about by changes in salt distribution across the plasma membrane.

It is a well-established fact that the loss of potassium during the volume regulatory shrinkage following osmotic swelling, results from an increased K^+ permeability. This was first shown by unidirectional flux measurements in duck erythrocytes (Kregenow, 1971a), in human erythrocytes (Poznansky & Solomon, 1972a, b), in frog oocytes (Sigler & Janáček, 1971b), in mouse leukaemic cells (Roti Roti & Rothstein, 1973) and in Ehrlich ascites tumour cells (Hendil & Hoffmann, 1974). Also, in cells with constitutively high Na^+ contents, such as dog erythrocytes, K^+ permeability increases with cell volume (Parker & Hoffman, 1976), but here a Ca^{2+}/Na^+ exchange system seems also to be involved (Parker et al., 1975).

In hypertonic solutions an increase in the permeability for the dominant external cation, Na^+, would function as a volume-regulating factor just as did the increase in K^+ permeability in hypotonic media. The apparent Na^+ permeability of the cell membrane is increased in hypertonic media for several cell types investigated. This has been shown in nucleated erythrocytes (Kregenow, 1971b; Cala, 1977) dog and cat erythrocytes (Sha'afi & Pascoe, 1973; Parker & Hoffman, 1976) and in Ehrlich ascites tumour cells (Hoffmann, 1978).

These early studies showed that these volume-controlling mechanisms utilise ouabain-insensitive transport processes to move Na^+ and K^+ across the membrane in a controlled fashion. At that time, net Na^+ and K^+ movements taking place when the sodium/potassium pump was blocked by ouabain, were believed to represent transport through 'leak' pathways. The 'leak' was considered to be a fixed process that was separate from exchange diffusion systems and the pump. Since these 'leak' transports during the volume-regulatory responses were dynamic and regulated, the static nature of the 'leak' had to be revised.

The study of osmoregulatory transport processes has therefore led to a new understanding of the Na^+ and K^+ 'leaks' in non-excitable cells. At least part of what was formerly defined as 'leak' demonstrates an unexpected responsiveness and transport capacity. In duck erythrocytes, Ehrlich ascites tumour cells and frog skin epithelium these transport processes appear to couple cation (Na^+ and/or K^+) movement to anion (Cl^-) movement (Cala, 1980; Kregenow, 1981; Hoffmann, Sjøholm & Simonsen, 1981; Ussing, 1982). In salamander erythrocytes, however, they can act as Na^+/H^+ exchanger (Kregenow, 1981). In dog erythrocytes they involve Ca^{2+}/Na^+ exchange, and in some mammalian cell lines a Ca^{2+}-gated, K^+ channel (Grinstein, DuPre & Rothstein, 1982; E. K. Hoffmann, unpublished).

This brief review focuses mainly on the volume-regulatory responses of Ehrlich ascites tumour cells. Relationships between these responses and those of other vertebrate cells are noted where appropriate. The reader is also referred to three more comprehensive reviews on the control of cell volume (Hoffmann, 1977; Macknight & Leaf, 1977; Rorive & Gilles, 1979); to a review dealing mainly with the responses of invertebrate cells to changes in external osmolarity (Gilles, 1979); and to a review concentrating on the well-described mechanisms involved in duck erythrocytes and in *Amphiuma* erythrocytes (Kregenow, 1981).

The list of vertebrate cells found to be capable of volume regulation in anisotonic media has grown, and in addition to those already mentioned includes human erythrocytes (Poznansky & Solomon, 1972*a*), proximal kidney tubule cells (Dellasega & Grantham, 1973), chick lymphocytes (Ben-Sasson *et al.*, 1975), toad tissues (Katz, 1978) and *Necturus* gallbladder epithelial cells (Spring & Persson, 1981).

Not all of these vertebrate cells respond in the same way to incubation in anisotonic solutions. Firstly, some cell types can correct their volume when incubated in hypotonic media but not in hypertonic media (Roti Roti & Rothstein, 1973; Hendil & Hoffmann, 1974; Ben-Sasson *et al.*, 1975). However, this is probably not a characteristic of these cells, since some of them can regulate their volume with modifications in the experimental conditions (Hoffmann *et al.*, 1981). Secondly, not all cells use only inorganic ions to adjust cell volume as do, for example, duck erythrocytes; other cells use amino acids as well (Fugelli, 1967; Lasserre & Gilles, 1971; Hoffmann & Hendil, 1976; Forster & Goldstein, 1979). Finally, although there is no convincing evidence that changes in metabolism modulate the volume-regulatory responses of most of these cells, changes in metabolism during the volume-regulatory response have been shown in Ehrlich ascites cells (Lambert & Hoffmann, 1982) and it has been demonstrated that the metabolic state can directly alter volume-induced transport in dog and cat erythrocytes (see Parker, 1977).

Control of cell volume at the steady state
Donnan equilibrium

The membrane is conceived of as being permeable to the solvent and to some, but not all, ions. The impermeable ions can be small ions or macromolecular polyelectrolytes. The equilibrium state in this situation is called the *Donnan equilibrium*. It is characteristic of the Donnan distribution that at equilibrium there is between the two compartments (1) an uneven (asymmetrical) distribution of the diffusible ions, (2) an electrical potential difference (the Donnan potential) and (3) an osmotic pressure difference. The following is a brief description of the Donnan distribution; a more complete treatment may be found in Sten-Knudsen (1978).

The Gibbs–Donnan rule states that the products of diffusible anions and cations on both sides of the membrane are equal at equilibrium. As a simple example let us consider the situation in Fig. 1 with the permeable cation (M^+) and anions (B^-) present both inside and outside of the cell and with an intracellular non-permeable anion (A^{n-}). From the Gibbs-Donnan rule we have (activity coefficients being neglected):

$$[M^+]_i \times [B^-]_i = [M^+]_o \times [B^-]_o \tag{1}$$

where the square brackets indicate the concentration and the subscripts the inner (i) and outer (o) compartments.

A state of electroneutrality must exist in each compartment, so that:

$$[M^+]_o = [B^-]_o \tag{2}$$

$$[M^+]_i = [B^-]_i + n[A^{n-}]_i \tag{3}$$

Now equation (1) can be rewritten:

$$[M^+]_i \times [B^-]_i = [M^+]_o^2 \tag{4}$$

If we remember that $[B^-]_i < [M^+]_i$ because of the presence of A^{n-}, then it follows from equation (4) that $[M^+]_i + [B^-]_i > 2[M^+]_o$. Since the total solute concentration inside the cell is equal to $[M^+]_i + [B^-]_i + [A^{n-}]_i$ and the solute concentration outside is equal to $2[M^+]_o$ then it can be seen that the total osmotic concentration will be greater inside than outside and there will be a positive osmotic pressure difference between the cell and its surroundings.

Water enters the cell and tends to level out the difference in osmotic pressure, but at the same time the ions will redistribute according to the Gibbs–Donnan rule. This will go on until all the diffusible ions are inside or until the cells burst (colloid osmotic lysis). I shall briefly demonstrate that colloid osmotic swelling in Ehrlich cells is avoided by a conventional 'pump and leak' system where active transport systems for Na^+ and K^+ render the cell membrane 'functionally' impermeable to Na^+ and K^+.

Fig. 1. A simple cell model. The cation M^+ and the anion B^- can penetrate the cell membrane. A^{n-} is an impermeable anion. Inside and outside are indicated by (i) and (o) respectively.

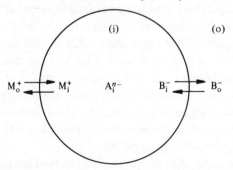

The 'pump and leak' concept

The pump and leak concept states that the membrane contains an active transport system for Na^+ and K^+ which renders the cell membrane functionally impermeable to these ions in that the active pump fluxes are equal to and oppositely directed to the leak fluxes. For discussion of this 'pump and leak' concept see Leaf (1959), Ussing (1960), Tosteson & Hoffman (1960) and Whittam (1964). From the 'pump and leak' concept it follows that an alteration in either the leakage permeabilities for the ions or the active transport rates will affect the total cation content of the cell, and thus the cell volume. If, for example, the leakage permeabilities are increased or the active transport rates decreased, the cell is no longer 'functionally' impermeable to cations. The Donnan effect thus will be expressed in a redistribution of ions and a consequent change in volume (see above). I shall demonstrate two examples of this:

If Ehrlich cells are chilled they take up electrolyte and water and swell. At 0 °C they swell to 1.5 times the original cell volume in about 100 min. When the cells are warmed again they regain their original volume in about 40 min (see Hoffmann, 1980). Another example is seen in Fig. 2, where the cells are transferred to high K^+ media. Here they also swell and the uptake of electrolytes accounts for the net water flux. Both these effects are expected

Fig. 2. Swelling of cells after transfer to Ringer solution where 150 mequiv l^{-1} of potassium has been substituted for sodium (K-Ringer), and shrinkage after resuspension in normal Ringer solution (Na-Ringer).

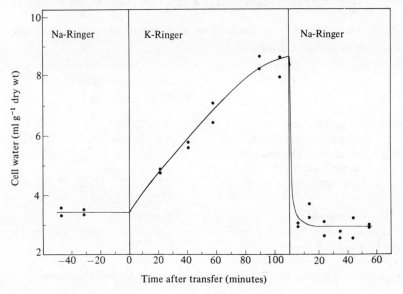

from the 'pump and leak' model and have also been described in many other systems.

Inhibition of the pump by ouabain, which generally produces a considerable swelling of cells, did not, however, affect the cell volume in Ehrlich cells very much. Table 1 shows the effect of ouabain on the cell volume and on the cellular concentration of K^+, ninhydrin-positive substances (a measure of non-protein amino acids, NPS), Na^+ and Cl^-. After a pre-incubation period of $1\frac{1}{2}$ hours, ouabain was added to a final concentration of 1 mM. During the next 50–80 minutes Na^+ increased by 23 mM over the initial concentration, K^+ decreased by 15 mM compared with the initial value and NPS decreased by 8 mM. The average cell volume showed a small, non-significant increase. As the concentration increase for Na^+ is 50% higher than the concentration decrease for K^+ the cells should have swelled after inhibition of the sodium/potassium pump, had the NPS not participated in the volume regulation. The transport of amino acids is linked to the transport of Na^+, and the intra- and extracellular concentrations of Na^+, K^+ and amino acids are interdependent (see e.g. Philo & Eddy, 1978). The amino acid changes in this experiment can at least partly result from the decrease in K^+ and Na^+ gradients.

Osmotic behaviour of animal cells

How is the volume of a cell affected by a change in the osmotic pressure of its environment?

The water permeability of most animal cells is extremely high. A noteworthy exception is fertilised fish eggs (Potts & Rudy, 1969; Loeffler & Løvtrup, 1970,

Table 1. *The effect of 1 mM ouabain on the concentrations of osmotically active solutes and on cell volume*

	Concentration change (mM)
Solutes	
Intracellular sodium	23 ± 2
Intracellular potassium	-15 ± 2
Intracellular chloride	3 ± 2
Intracellular NPS[a]	-8 ± 1
Total intracellular solutes	3 ± 4
	Volume change (%)
Cell volume	4.8 ± 1.3

From Hoffmann & Hendil (1976).
[a] NPS, ninhydrin-positive substances.
Samples were taken between 50 and 80 minutes after addition of ouabain.

where water permeability can be exceedingly low (Krogh & Ussing, 1937; Potts & Eddy, 1973; Riis-Vestergaard, 1982). Values of the hydraulic (osmotic) permeability for various cells can be found in Hoffmann (1977), Dick (1966) and House (1974).

Permeabilities to the predominant ions K^+, Na^+ and Cl^- are generally much lower than the water permeability. In the case of the erythrocyte, Na^+ and K^+ permeabilities are more than seven orders of magnitude lower and Cl^- permeability around five orders of magnitude lower than the water permeability. Other cells are, however, much more permeable to cations: for example, Ehrlich ascites cells have Na^+, K^+ and Cl^- permeabilities all five orders of magnitude lower than the water permeability (see Table 2).

The difference between ion and water permeabilities is still so large that the membrane of Ehrlich cells can be regarded as semipermeable. If a cell behaves as a perfect osmometer with a semipermeable membrane, its volume should depend on the osmotic pressure of the surrounding media according to the Boyle–Van't Hoff law:

$$V = V_0^{H_2O} \frac{\pi^o}{\pi} + b \tag{5}$$

where π^o and π are the isotonic external osmotic pressure and the actual osmotic pressure, respectively; V is the cell volume; b the non-solvent volume; and $V_0^{H_2O}$ the measured volume of cell water at isotonic external osmotic pressure. In most cells the degrees of swelling and shrinking are smaller than for a perfect osmometer. Thus the volume V is given by:

$$V = RV_0^{H_2O} \frac{\pi^o}{\pi} + b \tag{6}$$

where R is Ponder's R-value. Hempling (1960) and Hendil & Hoffmann (1974) observed a linear relationship between cell volume and osmolarity for Ehrlich cells transferred to hypertonic media, and from their data R-values of 0.8 and 0.9 respectively can be calculated (see Hoffmann, 1977). In a perfect

Table 2. *The membrane conductive permeability to K^+, Na^+, Cl^- and the osmotic permeability to water in Ehrlich ascites tumour cells*

Permeability constant	Value ($\times 10^{-8}$ cm s^{-1})
P_K	10.4
P_{Na}	4.4
P_{Cl}	3.9
$P_{osm(H_2O)}$ [a]	8.9×10^5

[a] From Hempling (1967); the other values are from Hoffmann *et al.* (1979).

osmometer R is unity. One of the few examples of such perfect osmotic behaviour is reported by Kwant & Seeman (1970) for erythrocyte ghosts freshly prepared and with membranes sealed at 37 °C.

In an attempt to explain this discrepancy various authors have considered the following factors: (1) leakage of ions from the cells to the hypotonic medium or from the hypertonic medium into cells, (2) elasticity of the cell membrane, (3) 'bound water', (4) concentration dependence of the osmotic coefficient of haemoglobin, and (5) decrease in the net charge on the haemoglobin molecule with increased concentration. For a discussion of these factors see Hoffmann (1977).

Cell volume regulation in anisotonic media

Like most animal cells Ehrlich cells can continue to regulate their volume when transferred to anisotonic media. Fig. 9 is a schematic drawing of the responses of Ehrlich ascites tumour cells incubated in hypertonic media (volume-regulatory increase, VRI) or hypotonic media (volume-regulatory decrease, VRD).

Volume-regulatory decrease

Fig. 3 shows the volume changes in a variety of cells following exposure to hypotonic media. It is seen that the responses of all these cells can be divided into two phases: an initial phase of osmotic swelling; and a second, more prolonged phase of cell shrinkage, the volume-regulatory decrease or VRD (Kregenow, 1971a). During the osmotic phase the cells swell more or less as perfect osmometers, as has been discussed in the previous section. This swelling is usually rapid, but there are variations caused by differences in permeability to water. During the volume-regulatory phase the volume gradually decreases towards a new steady-state value that is usually slightly above the original. Thus, the readjustment is not perfect. The time needed for the volume regulation varies in different cell types from around 5 minutes up to about 6 hours. In bathing media of lower osmolarity, the volume readjustment is more prolonged, although it initially proceeds more rapidly (Kregenow, 1971a; Hendil & Hoffmann, 1974).

Fig. 9 shows that when Ehrlich cells are enlarged, a ouabain-insensitive transport process is activated that controls the egress of potassium chloride from the cell (Hendil & Hoffmann, 1974; Hoffmann, 1978). Water accompanies the salt, causing the cells to shrink and return to their original isotonic volume in 10–30 minutes (see Figs. 3 and 4). Several other cell types, e.g. mouse leukaemic cells (Roti Roti & Rothstein, 1973), human lymphocytes (Ben-Sasson et al., 1975), duck erythrocytes (Kregenow, 1971, 1974) and frog skin epithelium (Ussing, 1982), have analogous responses involving a potassium loss.

Such adjustment can be seen in Fig. 4 which shows the loss of water, K^+ and Cl^- from the cells during VRD. Summarised results for K^+ at several osmolarities are given in Hendil & Hoffmann (1974). The Na^+ content does not change significantly during VRD. This is also found in duck erythrocytes (Kregenow, 1971a). It is obvious (see Fig. 4) that when cells undergo VRD, more K^+ leaves the cell than Cl^-. The route taken by Cl^- in Ehrlich cells seems to be divided into two steps. In the first step Cl^- enters or leaves the cell with

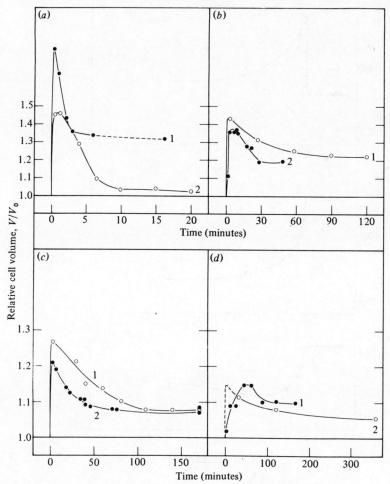

Fig. 3. Volume regulation of various cell types in hypotonic media. The relative cell volume (V/V_0) as a function of time after the cells were transferred to a hypotonic medium is shown. (a): 1, renal tubule cells (Dellasega & Grantham, 1973); 2, L 5178 Y cells (Roti Roti *et al.*, 1973). (b): 1, fish erythrocytes (Schmidt-Nielsen, 1975); 2, frog skin (MacRobbie & Ussing, 1961). (c): 1, crab muscle fibres (Lang & Gainer, 1969); 2, Ehrlich cells (Hendil & Hoffmann, 1974). (d): 1, frog oocytes (Sigler & Janáček, 1971a); 2, human erythrocytes (Poznansky & Solomon, 1972).

an equivalent amount of cation in what is effectively an electroneutral translocation of salt. In step two, Cl^-, having crossed the membrane, redistributes itself through passive pathway(s) associated with the anion exchanger or the anion/cation co-transporter. It is likely that OH^- (or more probably HCO_3^-) exchanges for Cl^-. In duck erythrocytes it is shown that the medium becomes alkaline during VRD (Kregenow, 1981), which supports this suggestion. It is, however, also possible that co-transport of sodium chloride into the cells is implicated at this stage of events since the redistribution

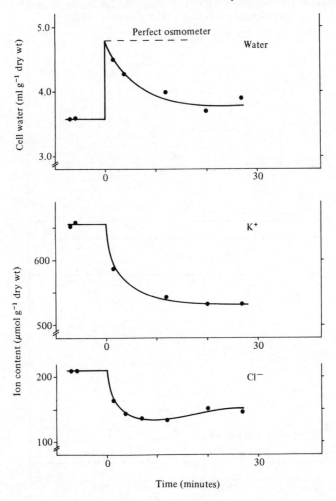

Fig. 4. Cell volume, K^+ and Cl^- content in Ehrlich ascites cells as a function of time after changes in medium osmolarity. Osmolarity was changed at $t = 0$ from 300 mosmol to 225 mosmol by dilution with distilled water.

of Cl⁻ is more pronounced when the co-transport is operating than when it is inhibited (see Fig. 7). It has been suggested that in frog skin epithelium the sodium chloride co-transport is induced during swelling (Ussing, 1982).

Cell-volume-dependent cation permeabilities. The decrease in K⁺ content of cells during VRD is dependent on the existence of an outwardly directed electrochemical K⁺ gradient and is brought about by a specific volume-induced increase in the membrane permeability to this cation. This is seen in Fig. 5, which shows K⁺ permeability (P_K) at various sodium chloride concentrations in the hypotonic range, together with controls in which sucrose had been added to a total osmolarity of 300 mosmol. P_K was calculated from the unidirectional K⁺ efflux in steady-state cells, using the Hodgkin & Katz expression (see e.g. Sten-Knudsen, 1978). The figure indicates that either cell volume or osmolarity *per se* triggers the changes in P_K.

The ratio of Na⁺ and K⁺ permeabilities may be calculated from the membrane potential, V_m, and the measured ion concentrations using the reduced Goldman equation (see Sten-Knudsen, 1978), assuming Cl⁻ to be in electrochemical equilibrium. The influence of cell volume on the ratio between

Fig. 5. The K⁺ permeability (P_K) of cells in media of varied osmolarity. Cells were incubated in media with different sodium chloride concentrations with (open circles) and without (filled circles) sucrose added to a total osmolarity of 300 mosmol, and steady-state fluxes measured with ⁴²K⁺. Asterisks indicate values from Hendil & Hoffmann (1974). (From Hoffmann, 1978.)

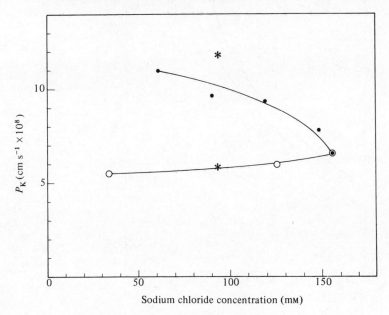

passive Na^+ and K^+ permeabilities is obvious from Fig. 6. From the measurement of P_K under the same experimental conditions it can be calculated that the increase in K^+ permeability is followed by a decrease in Na^+ permeability. Thus, the increased 'leak' pathway can only be used by K^+. The same response of cation permeabilities in hypotonic solution has been shown for dog erythrocytes (Parker & Hoffman, 1976).

K^+/Cl^- co-transport, K^+/H^+ exchange or Ca^{2+}-gated K^+ channel? If analysed separately from Cl^-, the volume dependent K^+ movement behaves as though the transport were a simple diffusion process. However, in duck erythrocytes (Kregenow, 1981; McManus & Haas, 1981), in genetically low K^+ sheep erythrocytes (Lauf & Theg, 1980; Dunham & Ellory, 1981), as well as in some fish erythrocytes (Lauf, 1982), the passive K^+ flux is dependent on the

Fig. 6. Effect of cell volume on the selective permeability ratio of the membrane for K^+ and Na^+ (P_K/P_{Na}). The sodium chloride concentration is kept constant at 75 mM (●) or 50 mM (○) and differences in osmolarity are obtained by addition of different amounts of sucrose. The ratio between the permeabilities is found from the reduced Goldmann equation: $V = (RT/F) \ln (P_{Na}/P_K)$, where V is the membrane potential, R the gas constant, T absolute temperature, F Faraday's number.

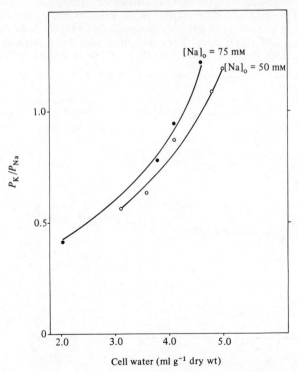

dominant anion and requires Cl⁻ (or Br⁻). This volume-sensitive K⁺ flux is inhibited by frusemide or bumetanide (McManus & Haas, 1981; Lauf, 1982), which are well-known inhibitors of co-transport systems in, for example, erythrocytes (Wiley & Cooper, 1974; Dunham, Stewart & Ellory, 1980) and Ehrlich cells (Geck *et al.*, 1978; Hoffmann *et al.*, 1981). For both of these the activation of a K⁺/Cl⁻ co-transport system is suggested as a component of VRD, and recently this has also been suggested for dog erythrocytes (J. C. Parker, personal communication).

In other cells, such as frog skin epithelium (Ussing, 1982) and Ehrlich ascites cells (Hoffmann, Sjøholm & Simonsen, 1983), frusemide or bumetanide do not inhibit the VRD. Fig. 7 (left) shows the loss of Cl⁻ during VRD with and without addition of bumetanide. Bumetanide has no effect on the initial loss of K⁺ and Cl⁻, whereas the secondary redistribution of Cl⁻ is slightly inhibited. In frog skin the loss of tissue K⁺ during exposure to diluted media is much larger in the presence of frusemide than in its absence. In the latter case, the co-transport carries sodium chloride into the cells and not potassium chloride out of the cells (Ussing, 1982). For Ehrlich cells and frog skin

Fig. 7. Cl⁻ content of Ehrlich ascites cells as a function of time after changes in medium osmolarity. Osmolarity was changed at the first $t = 0$ from 300 mosmol to 225 mosmol by dilution with distilled water and at the second $t = 0$ from 225 mosmol to 300 mosmol by addition of the salts present in the normal Ringer solution. Bumetanide (25 μM) was added at $t = 0$ in both cases. (Values from Hoffmann *et al.*, 1983.)

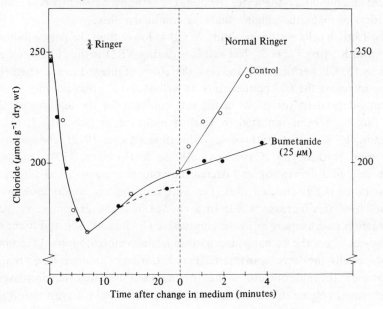

epithelium it is thus more likely that swelling increases separate 'leak' pathways.

In Ehrlich cells and in human lymphocytes evidence has been presented to support the idea that an elevation in the concentration of cytoplasmic Ca^{2+} that is triggered by cell swelling leads to an increase in the plasma membrane permeability to K^+. For example, it has been shown that quinine, an inhibitor of the Ca^{2+}-activated pathway for K^+ in other systems (Armando-Hardy et al., 1975), blocks the VRD in Ehrlich cells (E. K. Hoffmann, unpublished). In lymphocytes the effects of Ca^{2+} may be mediated by calmodulin (Grinstein, DuPre & Rothstein, 1982). Observations on *Amphiuma* erythrocytes suggest that energy within a cation gradient can be transferred to a pH gradient. This may indicate that a K^+/H^+ exchange is activated during VRD in parallel with a Cl^-/OH^- exchange or a Cl^-/HCO_3^- exchange (Cala, 1980; Kregenow, 1981).

Cell-volume-dependent anion transport. The net permeability to Cl^- is in the range 1×10^{-8} to 4×10^{-8} cm s^{-1} in Ehrlich cells (Hoffmann et al., 1979), in human erythrocytes (Dalmark, 1976; Hunter, 1977; Knauf et al., 1977) and in dog erythrocytes (Parker, Castranova & Goldinger, 1977). In addition a tightly coupled, electrically silent anion exchange mechanism is present in Ehrlich cells (Hoffmann et al., 1979) and in all erythrocytes (Wieth et al., 1974). In Ehrlich cells 95% of the Cl^- transport is mediated by the exchange diffusion (Hoffmann et al., 1979), whilst in *Amphiuma* erythrocytes (Lassen et al., 1978) and human erythrocytes (Brahm, 1977) the exchange flux is at least four orders of magnitude higher than the conductive flux.

In the Ehrlich cells net permeability to Cl^- is lower than the permeability to Na^+ and K^+ (see Table 2). Net salt loss during VRD is thus likely to be limited by the Cl^- permeability and it is, therefore, of interest to see whether swelling increases the Cl^- permeability as well as the K^+ permeability.

Volume-dependent variations in the rate constant for the unidirectional anion flux have been reported in Ehrlich ascites cells (see Fig. 8; also Hoffmann, 1978), in cat erythrocytes (Sha'afi & Pascoe, 1972), and in dog and human erythrocytes (Castranova, Weise & Hoffman, 1979). In the Ehrlich cells both increasing and decreasing volume are associated with an increase in the steady-state Cl^- flux (Fig. 8). Unfortunately, we do not know how much of this increase is due to a change in anion exchange, in the anion/cation co-transport or in the conductive Cl^- flux. In dog and human erythrocytes where the exchange mechanism is much more dominant than in the Ehrlich cells, the decrease in internal concentration of anions is larger than the increase in the rate constant, so that the anion self-exchange flux decreases with increased volume (Castranova et al., 1979). This could suggest that it is not the anion exchange which increases with increased volume in Ehrlich cells, but more likely the conductive flux.

Fig. 9 summarises the mechanisms involved in the VRD.

The role of amino acids and of taurine

A great relative decrease in the concentration of ninhydrin-positive substances occurs during VRD in Ehrlich cells. This decrease accounts for approximately 30% of the total decrease in osmotically active substances (Hoffmann & Hendil, 1976). The majority of the free amino acid pool is made up of non-essential amino acids such as alanine, glycine, glutamic acid, proline and aspartic acid. These make a major contribution to the intracellular adjustment during VRD, as can be seen from Table 3 where the concentrations of the five main amino acids and of taurine are shown. Similar results have been found for all invertebrate phyla studied so far, as well as for some vertebrates (for a review see Gilles, 1979).

A priori such regulation of the amino acid pool may be controlled either by a change in the rates of synthesis and degradation of these compounds,

Fig. 8. Volume dependence of the rate constant for $^{36}Cl^-$ efflux measured as unidirectional steady-state flux. The sodium chloride concentration is in all cases 75 mM, and differences in osmolarity are obtained by addition of different amounts of sucrose.

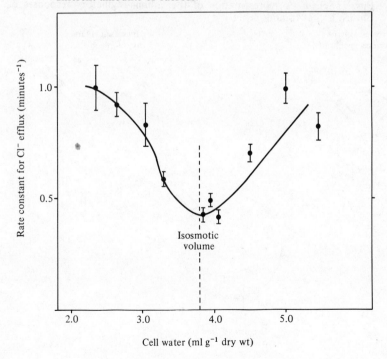

by the modification of protein metabolism, or by a change in the rate of transport through the cell membrane. In Ehrlich cells we have found an increase in the oxidative catabolism of both alanine and glycine under hypo-osmotic conditions. In the case of alanine this increased degradation accounts for 33% of the decrease in cellular alanine content, while the degradation of glycine and taurine plays no significant role. Changes in the rate of protein turnover do not seem to be involved (Lambert & Hoffmann, 1982).

Of the three possibilities proposed above, changes in taurine and amino acid transport in hypotonic media are most important. In Ehrlich cells, transport of taurine and glycine has been further analysed. It is found that a 30% increase in cell volume causes an increase in the diffusional taurine permeability of 7 times and in the diffusional glycine permeability of 1.5 times (Hoffmann & Lambert, 1983). The same increase in volume doubles the K^+ permeability (Hoffmann, 1977). The maximal flux for the Na^+-coupled glycine transport is decreased under the same conditions (Hoffmann & Lambert, 1983). Thus an increase in passive permeability to certain amino acids and a decrease in the Na^+-dependent amino acid influx results in a leakage of amino acids from the cellular pool to the external medium during VRD in Ehrlich cells.

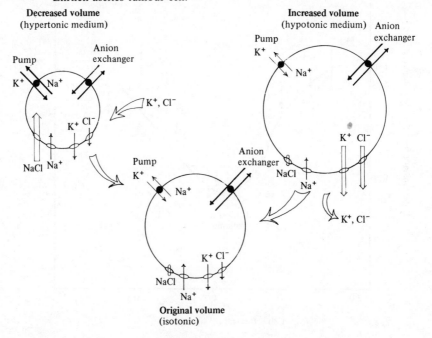

Fig. 9. Schematic representation of the volume-regulatory responses of Ehrlich ascites tumour cell.

Volume-regulatory increase

Fig. 9 shows the response of Ehrlich cells incubated in hypertonic media (VRI) as well as in hypotonic media (VRD).

Ehrlich ascites cell which have reached a steady state upon pre-incubation in hypotonic media (Hoffmann, 1978) shrink when transferred to a standard medium (now hypertonic compared with the cells) but return to their original volume within 10 minutes with an associated net uptake of potassium chloride (see Fig. 10). Osmotic shrinkage stimulates an otherwise quiescent ouabain-insensitive Na^+ and Cl^- transport system.

The assumption that the primary process is a coupled uptake of Na^+ and Cl^- followed by a replacement of Na^+ for K^+ by the sodium/potassium pump is supported by the following findings:

(1) Frusemide (1 mM), which is known to inhibit co-transport systems (Geck *et al.*, 1978), reduces water uptake and Cl^- uptake by 74%. Bumetanide (25 μM), which is shown to inhibit K^+/Cl^- co-transport in Ehrlich cells (Aull, 1981), reduces water uptake and Cl^- uptake by 88% (see Fig. 11).

(2) On replacing Cl^- with NO_3^-, net K^+ and water uptake are reduced by 60% (see Fig. 12).

(3) In media where Na^+ is replaced by choline (at 5 mM K^+), water and potassium chloride uptake are abolished. They are also strongly inhibited in media where K^+ is substituted for Na^+. The cells are hyperpolarised in choline medium but depolarised in K^+ medium. Thus, the net Cl^- uptake seems to be directly dependent on the Na^+ concentration (Hoffmann *et al.*, 1981, 1983).

Table 3. *Composition of the free amino acid pool in cells suspended in hypotonic (225 mosmol) and isotonic (300 mosmol) media*

	Concentration (mmol per litre cells)	
	300 mosmol ($n = 7$)	225 mosmol ($n = 8$)
Taurine	16.3 ± 2.1	7.6 ± 0.7
Aspartic acid	1.4 ± 0.2	1.0 ± 0.2
Glutamic acid	2.4 ± 0.3	1.6 ± 0.2
Proline	2.6 ± 0.4	2.1 ± 0.2
Glycine	10.3 ± 0.3	7.6 ± 0.4
Alanine	11.1 ± 1.1	7.0 ± 0.4
Total ninhydrin-positive substances	64.2 ± 1.1	42.4 ± 1.0

Values from Hoffmann & Hendil (1976).
n is the number of experiments. Values are given $\pm$S.E.M.

(4) K⁺ influx is stimulated and increases by approximately the same amount as the Cl⁻ influx (Fig. 13).

Water and potassium chloride uptake are insensitive to DIDS (200 μM), a reagent which inhibits the self-exchange flux of ^{36}Cl⁻ by 85% (Sjøholm, Hoffmann & Simonsen, 1981). The net Cl⁻ flux is more than 10 times higher

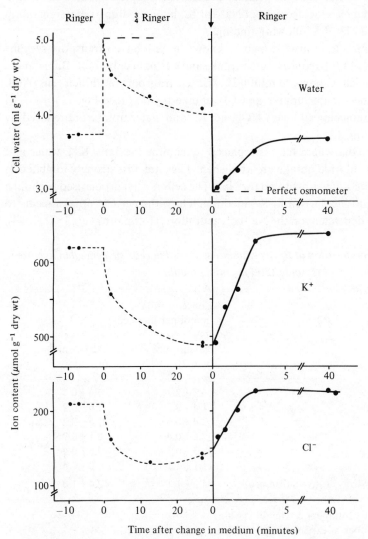

Fig. 10. Cell volume, K⁺ and Cl⁻ content in Ehrlich ascites cells as a function of time after changes in medium osmolarity. Osmolarity was changed at the first $t = 0$ from 300 mosmol to 225 mosmol by dilution with distilled water and at the second $t = 0$ from 225 mosmol to 300 mosmol by addition of the salts present in the normal Ringer solution.

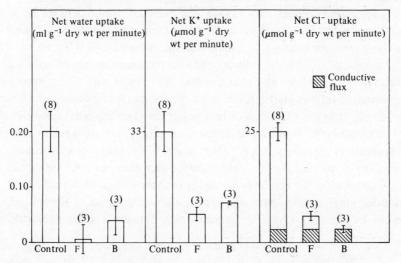

Fig. 11. Effect of the co-transport inhibitors frusemide (F: 1 mM) and bumetanide (B: 25 μM) on the net uptake of water, K^+ and Cl^- during volume-regulatory decrease. The experimental conditions were identical to those in Fig. 7. Values are given ±S.E.M. with the number of separate experiments in brackets. (Values from Hoffmann et al., 1983.)

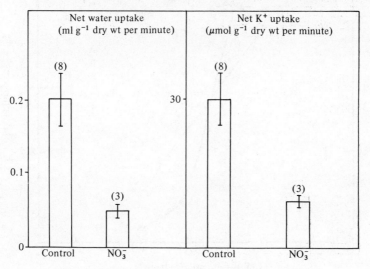

Fig. 12. Effect of replacing Cl^- with NO_3^- on the net uptake of water and K^+ during volume-regulatory increase. The experimental conditions were identical to those in Fig. 7. Values are given ±S.E.M. with the number of separate experiments in brackets. (Values from Hoffmann et al., 1983.)

than expected from the conductive Cl⁻ permeability previously reported (Hoffmann et al., 1979).

Taken together these results demonstrate the presence of an electrically silent, bumetamide- and frusemide-sensitive co-transport of Cl⁻ and Na⁺. A co-transport system for Na⁺, K⁺ and Cl⁻ in Ehrlich ascites cells has recently been suggested by Geck et al. (1980). The eventual requirement for K⁺ in the VRI transport process is not clear in the Ehrlich cells. Removing K⁺ from the bathing medium blocks Cl⁻ uptake but not immediately. If we assume that the uptake of Cl⁻ by the cells depends on the maintenance of an inwardly directed Na⁺ gradient and that the Na⁺ that enters with Cl⁻ is constantly removed, it follows that inhibition of the sodium/potassium pump should gradually stop the Cl⁻ uptake. This might be what happens, since removal of extracellular K⁺ is known to inhibit the sodium/potassium pump. Another possibility is that there is a K⁺, Na⁺ and Cl⁻ co-transport system but that recycling of cellular K⁺ for a short period can occur in a K⁺-free medium.

In conclusion, the results presented here show that co-transport of sodium chloride plays an important role in the volume regulation of Ehrlich cells. It could, however, also be a mechanism which transports sodium chloride and

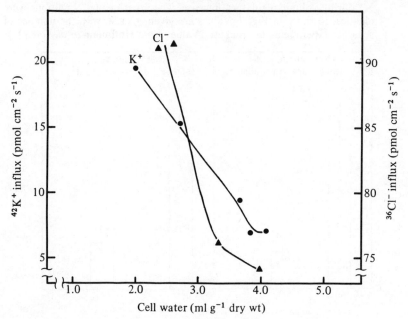

Fig. 13. Effect of cell volume on K^+ influx (measured as $^{42}K^+$ steady-state exchange flux) and on Cl^- influx (measured as $^{36}Cl^-$ steady-state exchange flux). The sodium chloride concentration is in all cases 75 mM, and differences in osmolarity are obtained by addition of different amounts of sucrose. Normal isotonic volume is around 3.8 ml per g dry wt.

potassium chloride together in the ratio 1:1, as has been proposed by Geck et al. (1980), and our present data do not permit a distinction between the two models. In duck erythrocytes a requirement for all three ions has been clearly demonstrated and it has been proposed that they cross the membrane as an entity (Kregenow, 1981). Evidence for a frusemide-sensitive Na^+/K^+ co-transport has been presented by Kregenow (1976, 1977, 1978), Schmidt & McManus (1977a, b, c) and McManus & Schmidt (1978). Evidence for the concept of Cl^-/cation co-transport has been summarised by Kregenow (1981).

Under physiological steady-state conditions the Ehrlich cells are relatively impermeable to Cl^- (Hoffmann et al., 1979) and there is very little co-transport of sodium chloride into the cells (Sjøholm et al., 1981; Hoffmann et al., 1981). This is also indicated by the finding that the cells retain their volume for relatively long periods of time after addition of the co-transport inhibitors frusemide or bumetanide (E. K. Hoffmann, unpublished). The unidirectional Cl^- flux in cells in isotonic media is also barely affected when Na^+ is replaced by choline, or the co-transport inhibitor bumetanide is added (Table 4). During VRI, however, the Na^+-dependent co-transport system seems to be activated and to play an important role. When the cells have attained their normal volume the co-transport mechanism is again inactivated. A similar activation of anion/cation co-transport in shrunken cells has recently been suggested for duck erythrocytes (Kregenow, 1981) and for epithelial cells (Ussing, 1982).

The trigger for activation of the Na^+/Cl^- co-transport is not clear. It cannot be only the cell volume which controls the activation of the co-transport system, since Ehrlich cells shrunken by addition of sucrose do not regulate their volume (Hendil & Hoffmann, 1974), although the sum of the chemical potentials for Na^+ and Cl^- still provides the necessary driving force for sodium chloride uptake. It is conceivable that cellular Cl^- concentration is a critical factor. Ussing (1982) has recently suggested that in frog skin the

Table 4. *The effect of bumetanide on steady-state chloride influx*

	Chloride influx (μmol g^{-1} dry wt min^{-1})
Control	44±2 (7)
Bumetanide (25 μM)	44±2 (3)
Bumetanide (100 μM)	45 (1)

From Hoffmann et al. (1983).
Values are ±S.E.M. with the number of experiments in brackets.

co-transport system is activated when cellular Cl⁻ concentration drops below a critical level. For Ehrlich cells the available evidence suggests that the intracellular Cl⁻ concentration has to be low, but that some additional factor connected with changes in cell volume is also involved in the activation of the anion/cation co-transport system.

Lis H. Christensen and Villy Rasmussen have given valued assistance in the preparation of the manuscript. Professor H. H. Ussing and Dr Lars Ole Simonsen are gratefully acknowledged for stimulating discussions and critical reading of the manuscript.

References

Armando-Hardy, M., Ellory, C. J., Ferreira, H. G., Fleminger, S. & Lew, V. L. (1975). Inhibition of the calcium-induced increase in the potassium permeability of human red blood cells by ouinine. *Journal of Physiology*, **250**, 32P–33P.

Aull, F. (1981). Potassium chloride cotransport in steady-state ascites tumor cells. Does bumetanide inhibit? *Biochimica et Biophysica Acta*, **643**, 339–45.

Ben-Sasson, S., Shaviv, R., Bentwich, Z., Slavin, S. & Doljanski, F. (1975). Osmotic behavior of normal and leukemic lymphocytes. *Blood*, **46**, 891–9.

Brahm, J. (1977). Temperature-dependent changes of chloride transport kinetics in human red cells. *Journal of General Physiology*, **70**, 283–306.

Cala, P. M. (1977). Volume regulation by flounder red blood cells in anisotonic media. *Journal of General Physiology*, **69**, 537–52.

Cala, P. M. (1980). Volume regulation by *Amphiuma* red blood cells. The membrane potential and its implications regarding the nature of the ion-flux pathways. *Journal of General Physiology*, **76**, 683–708.

Castranova, V., Weise, M. J. & Hoffman, J. F. (1979). Anion transport in dog, cat, and human red cells. Effects of varying cell volume and Donnan ratio. *Journal of General Physiology*, **74**, 319–34.

Dalmark, M. (1976). Chloride in human erythrocyte. Distribution and transport between cellular and extracellular fluids and structural features of the cell membrane. *Progress in Biophysics and Molecular Biology*, **31**, 145–64.

Dellasega, M. & Grantham, J. J. (1973). Regulation of renal tubule cell volume in hypotonic media. *American Journal of Physiology*, **224**, 1288–94.

Dick, D. A. T. (1966). *Cell Water*. London: Butterworth.

Dunham, P. B. & Ellory, J. C. (1981). Passive potassium transport in low potassium sheep red cells: dependence upon cell volume and chloride. *Journal of Physiology*, **318**, 511–30.

Dunham, P. B., Stewart, G. W. & Ellory, J. C. (1980). Chloride-activated passive potassium transport in human erythrocytes. *Proceedings of the National Academy of Sciences, USA*, **77**, 1711–15.

Forster, R. P. & Goldstein, L. (1979). Amino acids and cell volume regulation. *Yale Journal of Biology and Medicine*, **52**, 497–515.

Fugelli, K. (1967). Regulation of cell volume in flounder (*Pleuronectes flesus*) erythrocytes accompanying a decrease in plasma osmolarity. *Comparative Biochemistry and Physiology*, **22**, 253–60.

Geck, P., Heinz, E., Pietrzyk, C. & Pfeiffer, B. (1978). The effect of furosemide on the ouabain-insensitive K^+ and Cl^- movement in Ehrlich cells. In *Cell Membrane Receptors for Drugs and Hormones*, ed. R. W. Straub & L. Bolis, pp. 301–7. New York: Raven Press.

Geck, P., Pietrzyk, C., Burckhardt, B.-C., Pfeiffer, B. & Heinz, E. (1980). Electrically silent cotransport of Na^+, K^+ and Cl^- in Ehrlich cells. *Biochimica et Biophysica Acta*, **600**, 432–47.

Gilles, R. (1979). Intracellular organic osmotic effectors. In *Mechanisms of Osmoregulation in Animals*, ed. R. Gilles. pp. 111–53. New York: Wiley.

Grinstein, S., DuPre, A. & Rothstein, A. (1982). Volume regulation by human lymphocytes. Role of calcium. *Journal of General Physiology*, **79**, 849–68.

Hempling, H. G. (1960). Permeability of the Ehrlich ascites tumor cell to water. *Journal of General Physiology*, **44**, 365–79.

Hempling, H. G. (1967). Application of irreversible thermodynamics to a functional description of the tumor cell membrane. *Journal of Cellular Physiology*, **70**, 237–56.

Hendil, K. B. & Hoffmann, E. K. (1974). Cell volume regulation in Ehrlich ascites tumor cells. *Journal of Cell Physiology*, **84**, 115–25.

Hoffman, J. F. (1958). Physiological characteristics of human red blood cell ghosts. *Journal of General Physiology*, **42**, 9–28.

Hoffmann, E. K. (1977). Control of cell volume. In *Transport of Ions and Water in Animals*, ed. B. J. Gupta, R. B. Moreton, J. L. Oschman & B. J. Wall, pp. 285–332. New York & London: Academic Press.

Hoffmann, E. K. (1978). Regulation of cell volume by selective changes in the leak permeabilities of Ehrlich ascites tumor cells. In *Osmotic and Volume Regulation*, Alfred Benson Symposium XI, ed. C. B. Jørgensen & E. Skadhauge, pp. 397–417. Copenhagen: Munksgaard.

Hoffmann, E. K. (1980). Cell volume regulation in mammalian cells. In *Animals and Environmental Fitness*, ed. R. Gilles, pp. 43–59. Oxford: Pergamon Press.

Hoffmann, E. K. & Hendil, K. B. (1976). The role of amino acids and taurine in isosmotic intracellular regulation in Ehrlich ascites mouse tumor cells. *Journal of Comparative Physiology*, **108**, 279–86.

Hoffmann, E. K. & Lambert, I. (1983). Amino acid transport and cell volume regulation in Ehrlich ascites tumour cells. *Journal of Physiology*, in press.

Hoffmann, E. K., Simonsen, L. O. & Sjøholm, C. (1979). Membrane potential, chloride exchange, and chloride conductance in Ehrlich mouse ascites tumour cells. *Journal of Physiology*, **296**, 61–84.

Hoffmann, E. K., Sjøholm, C. & Simonsen, L. O. (1981). Anion–cation co-transport and volume regulation in Ehrlich ascites tumour cells. *Journal of Physiology*, **319**, 94P–95P.

Hoffmann, E. K., Sjøholm, C. & Simonsen, L. O. (1983). Na^+, Cl^- co-transport in Ehrlich ascites tumour cells activated during volume regulation. *Journal of Membrane Biology*, in press.

House, C. R. (1974). In *Water Transport in Cells and Tissues*, ed. H. Davson, A. D. M. Greenfield, R. Whittam & G. S. Brindley. London: Edward Arnold.

Hunter, M. J. (1977). Human erythrocyte anion permeabilities measured under conditions of net charge transfer. *Journal of Physiology*, **268**, 35–49.

Katz, U. (1978). Ionic and volume regulation in selected tissues of the euryhaline toad *Bufo viridis*. In *Osmotic and Volume Regulation*, ed. C. B. Jørgensen & E. Skadhauge, pp. 379–91. Copenhagen: Munksgaard.

Knauf, P. A., Fuhrmann, G. F., Rothstein, S. & Rothstein, A. (1977). The relationship between anion exchange and net anion flow across the human red blood cell membrane. *Journal of General Physiology*, **69**, 363–86.

Kregenow, F. M. (1971*a*). The response of duck erythrocytes to nonhemolytic hypotonic media. Evidence for a volume-controlling mechanism. *Journal of General Physiology*, **58**, 372–95.

Kregenow, F. M. (1971*b*). The response of duck erythrocytes to hypertonic media. *Journal of General Physiology*, **58**, 396–412.

Kregenow, F. M. (1974). Functional separation of the Na–K exchange pump from the volume controlling mechanism in enlarged duck red cells. *Journal of General Physiology*, **64**, 393–412.

Kregenow, F. M. (1976). Cell volume control. In *Water Relations in Membrane Transport in Plants and Animals*, ed. A. M. Jungries, T. K. Hodges, A. Kleinzeller & S. G. Schultz, pp. 291–302. New York & London: Academic Press.

Kregenow, F. M. (1977). Transport in avian red cells. In *Membrane Transport in Red Cells*, ed. J. C. Ellory & L. Lew, pp. 383–426. New York & London: Academic Press.

Kregenow, F. M. (1978). An assessment of the co-transport hypothesis as it applies to the norepinephrine and hypertonic responses. In *Osmotic and Volume Regulation*, ed. C. B. Jørgensen & E. Skadhauge, pp. 379–91. Copenhagen: Munksgaard.

Kregenow, F. M. (1981). Osmoregulatory salt transporting mechanisms: control of cell volume in anisotonic media. *Annual Review of Physiology*, **43**, 493–505.

Krogh, A. & Ussing, H. H. (1937). A note on the permeability of trout eggs to D_2O and H_2O. *Journal of Experimental Biology*, **14**, 35–7.

Kwant, W. O. & Seeman, Ph. (1970). The erythrocyte ghost is a perfect osmometer. *Journal of General Physiology*, **55**, 208–19.

Lambert, I. & Hoffmann, E. K. (1982). Amino acid metabolism and protein turnover under different osmotic conditions in Ehrlich ascites tumor cells. *Journal of Molecular Physiology*, **2**, 273–86.

Lang, M. A. & Gainer, H. (1969). Volume control by muscle fibers of the blue crab. *Journal of General Physiology*, **53**, 323–41.

Lassen, U. V., Pape, L. & Vestergaard-Bogind, B. (1978). Chloride conductance of the *Amphiuma* red cell membrane. *Journal of Membrane Biology*, **39**, 27–48.

Lasserre, P. & Gilles, R. (1971). Modification of the amino acid pool in the parietal muscle of two euryhaline teleosts during osmotic adjustment. *Experientia*, **27**, 1434–5.

Lauf, P. K. (1982). Evidence for chloride dependent potassium and water transport induced by hyposmotic stress in erythrocytes of the marine teleost *Opsanus tau*. *Journal of Comparative Physiology*, in press.

Lauf, P. K. & Theg, B. E. (1980). A chloride dependent K^+ flux induced by N-ethylmaleimide in genetically low K^+ sheep and goat erythrocytes. *Biochemical and Biophysical Research Communications*, **92**, 1422–8.

Leaf, A. (1959). Maintenance of concentration gradients and regulation of cell volume. *Annals of the New York Academy of Sciences*, **72**, 396–404.

Loeffler, C. A. & Løvtrup, S. (1970). Water balance in the salmon eggs. *Journal of Experimental Biology*, **52**, 291–8.

Macknight, A. D. C. & Leaf, A. (1977). Regulation of cellular volume. *Physiological Reviews*, **57**, 510–73.

McManus, T. J. & Haas, M. (1981). Catecholamine stimulation of K/K (K/Rb) exchange in duck red cells. *Federation Proceedings*, **40**, 484.

McManus, T. J. & Schmidt, W. F. III (1978). Ion and co-ion transport in avian red cells. In *Membrane Transport Processes*, vol. 1, ed. J. F. Hoffman, pp. 79–106. New York: Raven Press.

MacRobbie, E. A. C. & Ussing, H. H. (1961). Osmotic behaviour of the epithelial cells of frog skin. *Acta Physiologica Scandinavica*, **53**, 348–65.

Parker, J. C. (1977). Solute and water transport in dog and cat red blood cells. In *Membrane Transport in Red Cells*, ed. C. Ellory & V. L. Lew, pp. 427–65. New York & London: Academic Press.

Parker, J. C., Castranova, V. & Goldinger, J. M. (1977). Dog red blood cells: Na and K diffusion potentials with extracellular ATP. *Journal of General Physiology*, **69**, 417–30.

Parker, J. C., Gitelman, H. J., Glosson, P. S. & Leonard, D. L. (1975). Role of calcium in volume regulation by dog red blood cells. *Journal of General Physiology*, **65**, 84–96.

Parker, J. C. & Hoffman, J. F. (1976). Influences of cell volume and adrenalectomy on cation flux in dog red blood cells. *Biochimica et Biophysica Acta*, **433**, 404–8.

Philo, R. D. & Eddy, A. A. (1978). Equilibrium and steady state models of the coupling between the amino acid gradient and the sodium electrochemical gradient in mouse ascites tumour cells. *Biochemical Journal*, **174**, 811–17.

Potts, W. T. W. & Eddy, F. M. (1973). The permeability to water of the eggs of certain marine teleosts. *Journal of Comparative Physiology*, **82**, 305–15.

Potts, W. T. W. & Rudy, P. P., Jr (1969). Water balance in the eggs of the Atlantic salmon *Salmo salar*. *Journal of Experimental Biology*, **50**, 223–37.

Poznansky, M. & Solomon, A. K. (1972a). Regulation of human red cell volume by linked cation fluxes. *Journal of Membrane Biology*, **10**, 259–66.

Poznansky, M. & Solomon, A. K. (1972b). Effect of cell volume on potassium transport in human red cells. *Biochimica et Biophysica Acta*, **274**, 111–18.

Riis-Vestergaard, J. (1982). Water permeation in plaice eggs (*Pleuronectes platessa* L.). *Acta Physiologica Scandinavica*, **114**, 26A.

Rorive, G. & Gilles, R. (1979). Intracellular inorganic osmotic effectors. In *Mechanisms of Osmoregulation in Animals*, ed. R. Gilles, pp. 83–109. New York: Wiley.

Roti Roti, L. W. & Rothstein, A. (1973). Adaptation of mouse leukemic cells (L5178Y) to anisotonic media. *Experimental Cell Research*, **79**, 295–310.

Schmidt, W. F. III & McManus, T. J. (1977a). Ouabain-insensitive salt and water movements in duck red cells. I. Kinetics of cation transport under hypertonic conditions. *Journal of General Physiology*, **70**, 59–81.

Schmidt, W. F. III & McManus, T. J. (1977b). Ouabain-insensitive salt and water movements in duck red cells. II. Norepinephrine stimulation of sodium plus potassium cotransport. *Journal of General Physiology*, **70**, 81–97.

Schmidt, W. F. III & McManus, T. J. (1977c). Ouabain-insensitive salt and water movements in duck red cells. III. The role of chloride in the volume response. *Journal of General Physiology*, **70**, 99–121.

Schmidt-Nielsen, B. (1975). Comparative physiology of cellular ion and volume regulation. *Journal of Experimental Zoology*, **194**, 207–20.

Sha'afi, R. I. & Pascoe, E. (1972). Sulfate flux in high sodium cat red cells. *Journal of General Physiology*, **59**, 155–66.

Sha'afi, R. I. & Pascoe, E. (1973). Further studies of sodium transport in feline red cells. *Journal of General Physiology*, **61**, 709–26.

Sigler, K. & Janáček, K. (1971a). The effect of non-electrolyte osmolarity on frog oocytes. I. Volume changes. *Biochimica et Biophysica Acta*, **241**, 528–38.

Sigler, K. & Janáček, K. (1971b). The effect of non-electrolyte osmolarity on frog oocytes. II. Intracellular potential. *Biochimica et Biophysica Acta*, **241**, 539–46.

Sjøholm, C., Hoffmann, E. K. & Simonsen, L. O. (1981). Anion–cation co-transport and anion exchange in Ehrlich ascites tumour cells. *Acta Physiologica Scandinavica*, **112**, 24A.

Spring, K. R. & Persson, B. E. (1981). Quantitative light microscopy and epithelial function. In *Epithelial Ion and Water Transport*, ed. A. D. C. MacKnight & J. P. Leader, pp. 15–21. New York: Raven Press.

Sten-Knudsen, O. (1978). Passive transport processes. In *Membrane Transport in Biology*, vol. 1, ed. G. Giebisch, D. C. Tosteson & H. H. Ussing, pp. 5–113. Berlin, Heidelberg & New York: Springer Verlag.

Tosteson, D. C. & Hoffman, J. F. (1960). Regulation of cell volume by active cation transport in high and low potassium sheep red cells. *Journal of General Physiology*, **44**, 169–94.

Ussing, H. H. (1960). Active and passive transport of the alkali metal ions. In *The Alkali Metal Ions in Biology*, ed. H. H. Ussing, P. Kruhöffer, J. Hess Thaysen & N. A. Thorn, p. 67. Berlin, Göttingen & Heidelberg: Springer Verlag.

Ussing, H. H. (1982). Volume regulation of frog skin epithelium. *Acta Physiologica Scandinavica*, **114**, 363–9.

Whittam, R. (1964). *Transport and Diffusion in Red Blood Cells*. Baltimore: Williams & Wilkins.

Wieth, J. O., Funder, J., Gunn, R. B. & Brahm, J. (1974). Passive transport pathways for chloride and urea through the red cell membrane. In *Comparative Biochemistry and Physiology of Transport*, ed. L. Bolis, K. Bloch, S. E. Luria & F. Lynen, pp. 317–37. Amsterdam: North-Holland.

Wiley, J. S. & Cooper, R. A. (1974). A furosemide-sensitive cotransport of sodium plus potassium in human red cell. *Journal of Clinical Investigation*, **53**, 745–55.

E. A. NEWSHOLME and J. M. PAUL

The use of *in vitro* enzyme activities to indicate the changes in metabolic pathways during acclimatisation

The energy required by tissues (for, for example, movement, biosynthesis, ion transport, maintenance of structures) is obtained at a molecular level by the hydrolysis of ATP to ADP plus phosphate. The re-synthesis of ATP occurs in the processes of fuel oxidation within the cell. Since ATP is not stored in the cell, any increase in the rate of energy demand by a cell must result in an immediate and precise increase in the rate of fuel oxidation to provide the necessary ATP. In addition to these rapid regulatory responses, many animals and plants are subject to longer term changes in both the capacity for energy production and its utilisation by virtue of seasonal changes in environmental temperature, food availability, reproductive requirements and behavioural demands. The modification of metabolic pathways of energy storage, production and utilisation with the progression of the seasons clearly enables organisms to take advantage of favourable conditions, as well as to survive particularly unfavourable conditions.

Changes in temperature will affect the kinetic properties of some enzymes more than others and this may lead to changes in the flux through different metabolic pathways. Since the fluxes through metabolic pathways are determined by a limited number of key enzymes the adaptation of intermediary metabolism probably reflects adjustments at these steps in the overall sequence either by substitution of these enzymes in the longer term by homologous isoforms (see Johnston, this volume), by changes in their cellular concentration (Sidell, this volume) or perhaps by other regulatory mechanisms (e.g. covalent modification). The measurement of the relative capacities of different metabolic pathways and the identification of enzymes whose activities can provide an indication of such capacities are of considerable importance in understanding the adaptation of intermediary metabolism during acclimatisation.

Various fuels (e.g. glucose, glycogen, fatty acids, triglyceride, amino acids) are available for oxidation by the individual tissues and these may follow different routes or pathways of metabolism. One very important question for metabolic biochemists and environmental physiologists is which fuel(s) is being used to provide most of the energy for a given tissue? Indeed knowledge of the maximum capacity of energy-producing pathways is required before the physiological importance of the pathway can be appreciated and, in some cases, before the physiological role of the tissue can be properly assessed.

Various experimental techniques for the precise measurement of flux through a pathway have been developed, largely over the past 30 years. A description of methods has been given elsewhere (Newsholme, Zammit & Crabtree, 1978; Newsholme, Crabtree & Zammit, 1980), but one of these, which depends on the measurement of the maximum activities of certain key enzymes, is described in detail in this article. This method has several advantages over others: it is very simple, it can be used for comparative studies of a large range of animals, and it can be used in humans (or any large animal) usually with minimum surgical intervention. Unfortunately, its very simplicity has led to its misapplication leading to a lack of confidence in the technique. For this reason, the theoretical and experimental bases for the choice of enzymes that can be used as flux indicators are described below, and this is followed by an example of how this approach can be applied to the tricarboxylic acid (TCA) cycle, an important energy-producing pathway under aerobic conditions in many muscles.

Theoretical basis for the use of maximum enzyme activities as indicators of maximum flux

Before the reasons behind the choice of enzyme can be understood, the difference between near-equilibrium and non-equilibrium reactions must be explained, and the meaning of the term flux-generating step must be appreciated (for detailed reviews see Newsholme & Crabtree, 1976, 1979, 1981).

Near-equilibrium, non-equilibrium and flux-generating reactions

Near-equilibrium and non-equilibrium reactions. Reactions in a metabolic pathway can be divided into two classes: those that are very close to equilibrium (near-equilibrium) and those that are far removed from equilibrium (non-equilibrium). A reaction in a metabolic pathway is non-equilibrium if the activity of the enzyme which catalyses the reaction is low in comparison with the activities of other enzymes in the pathway, so that the concentration of substrate(s) of the reaction is maintained high whereas that of the product is maintained low. Consequently, the rate of the reverse component (V_r) of the reaction is very much less than the rate of the forward component (V_f). In the following example, the rate in the forward direction is 1000-fold greater than the rate in the reverse direction:

$$S \to A \underset{0.01}{\overset{10.01}{\rightleftharpoons}} B \to P$$

hence the reaction is non-equilibrium.

A reaction is near-equilibrium if the catalytic activity of the enzyme is high in relation to the activities of other enzymes in the pathway, so that the rates of the forward and the reverse components of the reaction are much greater than the overall flux. In the following example, the difference between the forward and reverse components is only 10% and the rate of the forward component is 10-fold greater than the flux:

$$\rightarrow A \underset{90}{\overset{100}{\rightleftharpoons}} B \rightarrow$$

It is also possible to explain the difference between near- and non-equilibrium reactions thermodynamically. The free energy change of a reaction (ΔG) can be calculated from a knowledge of the ratio of the concentration of product to the concentration of substrate (when measured in the living cell or tissue, and this is known as the mass action ratio, or Γ), and the equilibrium constant, K_{eq}, of the reaction. Thus

$$\Delta G = -RT \ln K_{eq} + RT \ln \Gamma$$

so that

$$\Delta G = RT \ln (K_{eq}/\Gamma)$$

where R is the gas constant and T is the absolute temperature (for the derivation of this equation see Crabtree & Taylor, 1979). It can be shown that there is a mathematical relationship between the thermodynamic and the kinetic interpretations of the equilibrium nature of reactions (see Newsholme et al., 1980) such that

$$\frac{V_f}{V_r} = \frac{K_{eq}}{\Gamma}$$

Newsholme & Crabtree (1976) have, on the basis of sensitivity in regulation, assumed that a value of the ratio K_{eq}/Γ or V_f/V_r greater than 5.0 (i.e. a ΔG value > 1.0 kcal or 4.2 kJ) indicates a non-equilibrium reaction, whereas a ratio of less than 5.0 (i.e. $\Delta G < 1.0$ kcal) indicates a near-equilibrium reaction.

The flux-generating reaction. If an enzyme catalyses a non-equilibrium reaction in a metabolic pathway and approaches saturation with its pathway-substrate (that substrate which represents the flow of matter through the pathway), and consequently the catalytic rate is independent of the substrate concentration, the reaction is known as the flux-generating step for the pathway. In other words, in the steady state, this reaction initiates a flux to which all the other reactions in the pathway must adjust. (Such a reaction must be saturated with its pathway-substrate since, if it were not, as the reaction proceeded the substrate concentration would decrease and this would decrease the rate of the reaction and hence the flux through the pathway; a steady state would then be impossible.) One important development from the concept of the flux-generating step is that it provides a physiologically useful definition of a metabolic pathway. A pathway is defined as a series, either short or long, of enzyme-catalysed reactions that is initiated by a flux-generating step and ends either with the loss of end-product(s) to the environment (e.g. carbon dioxide and water) or to a metabolic sink (such as a storage product), or in a reaction that precedes another flux-generating step. Previously there had not been a specific definition of a pathway. The current article concerns the maximum flux through such a pathway and the further biochemical and physiological interpretations based on the magnitude of this flux.

Near-equilibrium, non-equilibrium and flux-generating reactions in relation to flux

Flux-generating reactions. For any pathway the maximum flux through that pathway must be dependent upon the activity of the enzyme that catalyses the flux-generating step. By definition, this enzyme is saturated with its pathway-substrate so, provided that the concentration of any second substrate approaches saturation, the maximum activity of this enzyme should provide a quantitative indication of the maximum flux through the pathway *in vivo*. However, there can be problems with the use of 'flux-generating' enzymes in this way. First, the enzyme may not be saturated with the second substrate, so when the enzyme is assayed *in vitro*, at saturating concentrations of both substrates, the activity could be considerably greater than the flux *in vivo*. For example, citrate synthase (EC 4.1.3.7) is probably saturated with its pathway-substrate, acetyl-CoA, but it is not saturated with oxaloacetate (see Rowan & Newsholme, 1979). Consequently, maximal *in vitro* activities of citrate synthase are considerably higher than the flux through the cycle (Alp *et al.*, 1976). Secondly, it follows from the definition of a metabolic pathway given above that the pathway may span more than one tissue, so the flux-generating step may be present in a different tissue from most of the reactions in that pathway. For example, in the immediate post-absorptive state, hepatic phosphorylase (EC 2.4.1.1) is the flux-generating step for glycolysis-from-glucose in muscle and probably other tissues (Newsholme & Crabtree, 1979). However, measurement of the activity of hepatic phosphorylase will not provide an indication of the maximum rate of glycolysis-from-glucose in any given muscle.

In contrast, phosphorylase in muscle is the 'flux-generating' enzyme for glycolysis-from-glycogen and its maximum activity provides a quantitative indication of the maximum capacity of glycolysis-from-glycogen, which usually represents anaerobic glycolysis in this muscle (see Crabtree & Newsholme, 1972a; Newsholme *et al.*, 1978).

Near-equilibrium reactions. It should be clear from the above discussion that for a reaction to maintain near-equilibrium status, the rate of the forward component of the reaction must be considerably greater than the flux even during conditions of maximum flux. In the assay of the enzyme *in vitro*, the activity is usually measured in the forward direction and at saturating concentrations of substrate. Hence, the maximum *in vitro* activity will be very much greater than the maximum flux.

In many studies the activities of 'near-equilibrium' enzymes have been used as indices of the flux through a pathway. For example, glyceraldehyde-3-phosphate dehydrogenase (EC 1.2.1.12) activity has been used to indicate the glycolytic capacity in muscle (see Beenakkers, 1969; Pette, 1966). However, the maximum activity of this enzyme may be an order of magnitude greater than the maximum flux through the pathway (see Crabtree & Newsholme, 1975). (The maximum glycolytic flux in the muscles of the locust, cockroach, honey bee and rat heart is 14, 15, 32 and 3.7 whereas reported activities of glyceraldehyde-3-phosphate dehydrogenase are 330, 100, 150 and 240 μmol min^{-1} g^{-1} fresh muscle, respectively.) The same problem applies to the use of lactate dehydrogenase (EC 1.1.1.27) as a quantitative index of glycolysis, β-hydroxybutyryl-CoA dehydrogenase (EC 1.1.1.157) as a quantitative index of fatty acid oxidation, and fumarase (fumarate hydratase, EC 4.2.1.2) or malate dehydrogenase (EC 1.1.1.37) as quantitative indices of the TCA cycle.

Non-equilibrium reactions. It is possible that some metabolic pathways contain non-equilibrium reactions that, under most conditions, are not completely saturated with substrate (i.e. they are not flux-generating steps), but that during periods of maximum flux may approach saturation. The maximum *in vitro* activities of such enzymes could, therefore, provide a quantitative index of the maximal flux. Although it may be possible, from a detailed knowledge of their regulatory properties, to identify such enzymes theoretically, experience has shown that experimental verification is essential.

Experimental justification of the use of 'non-equilibrium' enzyme activities to indicate rate of fuel utilisation

The experimental proof that the activity of a 'non-equilibrium' enzyme can be used to give a quantitative indication of maximum flux depends on a comparison of the maximum *in vitro* activity of the enzyme with the measured or calculated maximum flux through the pathway. It is preferable to use more than one tissue and more than one animal in this investigation. The major difficulty in the approach is obtaining information on the maximum flux in the intact tissues. For fuel utilisation in muscle, such information can be obtained in several ways. First, the rate of fuel utilisation can be calculated from oxygen uptake of the working muscle (from the equation for fuel oxidation – this is used in obtaining the data in Tables 1 and 2). Secondly, if the assumption is made that most of the energy produced by the muscle will be utilised by the contractile process, the maximal *in vitro* activity of myofibrillar adenosine triphosphatase (EC 3.6.1.3) should indicate the maximal ATP requirement of the muscle. Hence the rate of fuel utilisation required to satisfy this rate of ATP hydrolysis can be calculated, assuming the usual stoichiometric relationship between metabolic pathways and ATP production. Thirdly, anaerobic muscle produces mainly lactate as an end-product (at least in vertebrates), so the rate of lactate production by working muscle *in vivo* or *in vitro* indicates the rate of glycogen utilisation. Fourthly, measurement of the rate of glucose (or glycogen) utilisation by the isolated working muscles or measurement of arteriovenous differences across an exercising muscle in the intact animal provides a direct method for indication of rates of fuel utilisation. Such experiments have been described elsewhere (see Crabtree & Newsholme, 1972*a*, 1975; Newsholme *et al.*, 1978, 1980) and are described for the TCA cycle below.

Oxoglutarate dehydrogenase activity as a quantitative index of aerobic metabolism in muscle

In an attempt to find a quantitative index of aerobic metabolism in muscle (i.e. flux through the TCA cycle and the electron transfer chain) which could be applied across the animal kingdom, a search for a rate-indicating enzyme of the TCA cycle was initiated several years ago. Studies on the enzymes of the TCA cycle have revealed several enzymes catalysing non-equilibrium reactions: citrate synthase, NAD^+-linked isocitrate dehydrogenase and oxoglutarate dehydrogenase. (There is good evidence that most of the enzymes of the electron transfer chain catalyse near-equilibrium reactions, so that they cannot be used.) Comparison of maximum TCA cycle fluxes, calculated on the basis of oxygen uptake data, has eliminated citrate synthase and NAD^+-linked isocitrate dehydrogenase (see Alp, Newsholme & Zammit, 1976). However, oxoglutarate dehydrogenase activities are similar to the flux through the TCA cycle, suggesting that this activity provides the best indication of the maximum flux through the cycle (Table 1). Hence a survey of the V_{max} values of this enzyme in muscle was carried out and the results provide a framework of knowledge about the maximum capacity of aerobic metabolism in these muscles (see Table 2). This knowledge can be used in conjunction with information concerning the maximum rates of other energy-producing pathways to discuss the relationship between the metabolism and the physiology in different muscles.

The emphasis of this volume is on an understanding of adaptation of animals to different conditions. In many instances, the adaptation of organisms to changes in environmental conditions requires some adjustment in the relative importance of the various pathways of intermediary metabolism. For example, if a change in salinity of the surrounding water required an increase in work by a marine organism to maintain osmotic balance, how would the major energy-producing pathways respond? Would there be an increase in the capacity of the most efficient energy-producing system, the TCA cycle? If this were the case, what fuel would be used to provide acetyl-CoA for the cycle: glucose, lipid or amino acid? A systematic study of the changes in key indicator enzymes during physiological adaptation to a changed environment would not only provide quantitative information on the metabolic response to the changed condition but might also provide further insight into the relationship between physiological requirements and metabolism. Thus a number of studies have compared the *in vitro* enzymatic activity of tissue extracts from temperature-acclimated animals. Hazel & Prosser (1974) and Shaklee *et al.* (1977) have reviewed this work and have concluded that those enzymes which display enhanced activities in cold-acclimated individuals are mainly associated with energy production (e.g. TCA cycle, glycolysis). The identification of the flux-generating enzymes in each pathway is of crucial importance in the interpretation of such information, as is the use of maximum activities of 'non-equilibrium' enzymes for assessing quantitative changes in capacity of pathways. It is clear that not all enzymes which have been studied catalyse non-equilibrium reactions in their respective pathways.

Nevertheless, the use of radiolabelled substrates has confirmed that intermediary metabolism is indeed restructured *in vivo* though the pattern varies from species to species and from tissue to tissue (see for example Stone & Sidell, 1982; Jones & Sidell, 1982).

A particularly close relationship between physiology and intermediary metabolism has been established by comparing muscles from different animals across the animal kingdom that have diverse physiology, and this is described in the following section, to illustrate the sort of information that can be obtained. This is then taken one stage further in a section in which the significance of different fuels for ultimate oxidation in the TCA cycle is briefly discussed in relation to their physiological requirements. These discussions may provide appropriate examples to help the interpretation of metabolic adjustments in response to adaptation to different conditions and of how changes in enzyme activities can be related both qualitatively and quantitatively to the flux through metabolic pathways and hence the physiological changes in the whole animal. The more controlled conditions of a physiological adaptation may permit some of the hypotheses described in these sections to be investigated more fully and in more detail.

The importance of the TCA cycle for energy provision in muscle

The maximal activities of oxoglutarate dehydrogenase agree well with estimates of maximum fluxes through the TCA cycle in both vertebrate and insect muscle (Table 1), so that this activity can be used as an index of the aerobic capacity in muscles from animals across the whole animal kingdom (Table 2). From these results, the muscles investigated can be divided arbitrarily into four groups: group 1 in which the maximum activity of oxoglutarate dehydrogenase is greater than 20, group 2 in which the activity lies between 5 and 20, group 3 in which the activity lies between 1 and 5 and group 4 in which the activity lies below 1 μmol min^{-1} g^{-1} at 25 °C. The muscles within these groups are shown in Table 3. It can be seen that group 1 comprises only insect flight muscles. This confirms, quantitatively, the view that this muscle is highly aerobic and has a high power output. The maximum power output of insect flight muscle is at least 10-fold greater than that of vertebrate muscle (Weis-Fogh, 1961) and this is also indicated in the present study, where the highest maximum activity of oxoglutarate dehydro-

Table 1. *Estimates of the flux through the TCA cycle from oxygen uptake data and maximum activities of oxoglutarate dehydrogenase*

Animal	Tissue	Calculated flux through TCA cycle (μmol·min^{-1}·g^{-1} at 25 °C)	Maximum activity of oxoglutarate dehydrogenase (μmol·min^{-1}·g^{-1} at 25 °C)
Honey bee	Flight muscle	53.0	49.7
Bumble bee (queen)	Flight muscle	46.1	45.9
Blowfly	Flight muscle	66.3	46.7
Cockroach	Flight muscle	22.3	48.6
Silver-Y moth	Flight muscle	24.8	25.9
Desert locust	Flight muscle	20.7	20.8
Trout	Red muscle	2.7	1.8
Pigeon	Pectoral muscle	4.3	3.2
Rat	Heart	6.3	6.8
	Soleus	2.2	2.4
Man	Quadriceps	1.0	1.2

The oxygen uptake data were collated from the references indicated and were converted to estimates of maximum flux through the TCA cycle at 25 °C using the conversion factors described by Crabtree & Newsholme (1972, 1975). The activities of oxoglutarate dehydrogenase are means given in Table 2. For references to oxygen uptake measurements see Paul (1979); for details of assay method for oxoglutarate dehydrogenase, see Cooney et al. (1981).

Table 2. *Maximal activities of oxoglutarate citrate synthase and total isocitrate dehydrogenase in muscles from invertebrates and vertebrates*

		Enzyme activities (μmol·min^{-1}·g^{-1} at 25 °C)		
Animal	Tissue	Oxoglutarate dehydrogenase	Citrate synthase	Total isocitrate dehydrogenase
Mollusca				
Common whelk (*Buccinum undatum*)	Radula retractor	4.7 (0.9) (4)	24.8	28.6
Common limpet (*Patella vulgata*)	Radula retractor	1.7 (0.6) (4)	26.8	109.5
Periwinkle (*Littorina littorea*)	Radula retractor	2.1 (1)	—	—
Insecta				
Lepidoptera				
Yellow underwing moth (*Noctua pronuba*)	Flight muscle	39.6 (9.5) (9)	296.0	179.0
Silver-Y moth (*Plusia gamma*)	Flight muscle	25.9 (1.2) (2)	352.0	129.0
Poplar hawk moth (*Laöthoe populi*)	Flight muscle	42.7 (15.2) (3)	430.0	109.0
Hymenoptera				
Common wasp (*Vespula vulgaris*)	Flight muscle	71.2 (11.0) (4)	332	116
German wasp (*Vespula germanica*)	Flight muscle	76.5 (9.5) (3)	—	—
Honey bee (*Apis mellifera*)	Flight muscle	49.7 (8.0) (5)	346	98

Table 2 (*cont.*)

Animal	Tissue	Enzyme activities (μmol·min^{-1}·g^{-1} at 25 °C)		
		Oxoglutarate dehydrogenase	Citrate synthase	Total isocitrate dehydrogenase
Bumble bee (*Bombus terrestris*)	Flight muscle			
Queen		45.9 (6.7) (4)	325	107
Worker		47.0 (7.3) (5)	382	153
Male		67.6 (14.3) (5)	349	116
Coleoptera				
Rosechafer (*Pachnoda ephippiata*)	Flight muscle	36.8 (7.1) (5)	561	83
Summer chafer (*Amphimallon solstialis*)	Flight muscle	57.5 (26.1) (5)	243	183
African dung beetle (*Heliocopris* sp.)	Flight muscle	23.3 (3.5) (7)	294	72
Great diving beetle (*Dysticus marginalis*)	Flight muscle	12.0 (3.8) (3)	26.8	3.8
Hemiptera				
Giant water bug (*Lethocerus cordofanus*)	Flight muscle	15.2 (1.6) (5)	244	104

Dictyoptera				
Cockroach (*Periplaneta americana*)	Flight muscle	48.6 (8.9) (7)	185	77
	Leg muscle	6.8 (0.2) (3)	—	—
Orthoptera				
Desert locust (*Schistocerca gregaria*)	Flight muscle	20.8 (4.6) (9)	242	62
African locust (*Locusta migratoria*)	Flight muscle	44.2 (4.2) (5)	—	—
Diptera				
Blowfly (*Calliphora vomitoria*)	Flight muscle	46.7 (11.0) (6)	418	180
Fleshfly (*Sarcophaga barbata*)	Flight muscle	32.8 (9.3) (12)	316	145
Housefly (*Musca domestica*)	Flight muscle	38.1 (10.7) (3)	345	103.2
Tsetse fly (*Glossina morsitans*)	Flight muscle	26.4 (0.8) (3)	74	29
Pisces				
Dogfish (*Scylliorhinus canicula*)	Heart	3.3 (0.7) (3)	46	53.4
	Red muscle	2.2 (0.6) (3)	35	31.4
	White muscle	< 0.1 (3)	1.7	0.6

Table 2 (*cont.*)

		Enzyme activities (μmol·min^{-1}·g^{-1} at 25 °C)		
Animal	Tissue	Oxoglutarate dehydrogenase	Citrate synthase	Total isocitrate dehydrogenase
Trout (*Salmo gairdneri*)	Heart	3.3 (0.9) (5)	38	101
	Red muscle	1.8 (0.4) (3)	50	125
	White muscle	< 0.1 (3)	5.2	9.2
Amphibia				
Frog (*Rana temporaria*)	Heart	3.5 (1.3) (5)	40	75
	Sartorius	0.1 (0.2) (6)	10	14
	Gastrocnemius	< 0.1 (3)		
African clawed toad (*Xenopus laevis*)	Heart	3.3 (1.0) (2)	7.0	5.5
	Sartorius	1.2 (0.1) (2)		
	Gastrocnemius	1.0 (0.5) (2)	12.0	12.0
Aves				
Chicken (*Gallus gallus*)	Heart	6.0 (0.8) (3)	51	59.7
	Pectoral muscle	< 0.1 (4)	6.8	3.3

Species	Tissue	Citrate synthase	Oxoglutarate dehydrogenase	Isocitrate dehydrogenase
Pigeon (*Columba livia*)	Heart	4.3 (1.2) (4)	127	74
	Pectoral muscle	3.2 (1.3) (4)	115	54.8
Pheasant (*Phasianus colchicus*)	Heart	4.7 (1.0) (2)	99.0	48.7
	Pectoral muscle	1.5 (1.0) (2)	14.4	3.5
Mammalia				
Laboratory mouse (*Mus musculus*, Balb/C)	Heart	8.9 (1.0) (5)	146	106
Laboratory rat (*Rattus norvegicus*)	Heart	6.8 (2.4) (7)	96	77
	Soleus	2.4 (0.2) (4)	—	—
	Diaphragm	2.4 (0.8) (5)	—	—
Guinea pig	Heart	7.1 (1)		
Rabbit (*Oryctolagus caniculus*)	Heart	9.0 (2.0) (4)	69	34.2
	Adductor longus	0.7 (0.2) (2)		
	Semitendinosus	0.9 (0.5) (4)		
Sheep (*Ovis aries*)	Heart	4.8 (0.8) (6)	58.0	21.6
Pig (*Sus scrofa*)	Heart	3.8 (0.4) (6)	50.0	13.6

Maximum activities of citrate synthase and isocitrate dehydrogenase (NAD$^+$- plus NADP$^+$-linked activities) are shown for comparison. These data are from Alp et al. (1976). Oxoglutarate dehydrogenase was assayed as described by Cooney et al. (1981) and activities are presented as medians, with the number of observations below this in parentheses. The figure adjacent to the medians, also in parentheses, is the value of S_w (see Dean & Dixon, 1951). This is an estimate of the standard deviation, which is obtained from the range.

genase in insect flight muscle (71.5 in wasp flight muscle) is approximately 8-fold higher than the highest in vertebrate muscle (9.0, in rabbit heart) and approximately 20-fold higher than that in vertebrate skeletal muscle (3.2 in pigeon pectoral muscle). The high power output of insect flight muscle is related to its small size (see Hill, 1950).

Group 2 includes some vertebrate hearts, the whelk radula retractor, and the flight muscles of the giant diving beetle (*Dytiscus*) and the giant water-bug. The latter two insects are mainly aquatic, but they are also good fliers, and flight is important for their dispersal. It is possible, since both these insects had been kept in captivity in water tanks for more than a month and fed, that their flying competence had been reduced, and although both insects could be made to fly in the laboratory they may not have possessed the full potential power output and the activity of oxoglutarate dehydrogenase may have been lower than expected for an insect flight muscle.

Vertebrate hearts and the whelk radular muscle are both muscles which are used continuously, or almost continuously, and they serve a vital function pumping blood around the body and feeding respectively (see below). Since it would be impossible to store sufficient fuel within the muscle for continuous contractile activity, such muscles require a continuous supply of exogenous fuel and, in vertebrates, this occurs via the blood system. Since the latter also supplies oxygen, provided a blood supply is available, aerobic metabolism of the fuels can occur and this is advantageous due to the higher efficiency of ATP production. The whelk is large for a gastropod (20 g weight as opposed to the limpet which weighs about 7 g) and needs a large intake of food to supply its energy requirements. As a carrion feeder, it possesses a sharp, rasping radula and a large retractor muscle to operate it (Owen, 1966). In order to provide sufficient food the radula retractor muscle may have to function continuously for long periods (e.g. 24 hours or more) and this requires efficiency of energy production and hence aerobic metabolism.

Of the vertebrates, only the birds and mammals have a complete double circulation of blood within the heart, i.e. a four-chambered heart. Fish possess a two-chambered heart, and the amphibian heart is three-chambered. The evolution of the double circulation allows a higher peripheral blood pressure to be developed by the heart and this requires a higher power output. This difference is reflected in the enzymology: thus hearts of birds and mammals have a higher activity of oxoglutarate dehydrogenase (3.8 to 9.0) than the hearts of fish and amphibians (3.3 to 3.5). The hearts of larger mammals such as the sheep, pig and cow have lower activities of this enzyme than hearts of the smaller mammals (Table 2), due probably to scale effects (see Hill, 1950).

Group 3, apart from containing the hearts of other vertebrates, also

Table 3. *Groups of muscles classified according to their* V_{max} *of oxoglutarate dehydrogenase* ($\mu mol \cdot min^{-1} \cdot g^{-1}$)

Group 1: > 20	Group 2: 5 to 20	Group 3: 1 to 5	Group 4: < 1
Flight muscles of:	Flight muscles of:	Hearts of:	Frog sartorius
Silver-Y moth	Giant diving beetle	Pigeon	Frog gastrocnemius
Yellow underwing moth	Giant water bug	Cow	Toad gastrocnemius
Poplar hawk moth	Whelk radular retractor	Pig	White muscles of:
Wasp (German and common)	Hearts of:	Dogfish	Dogfish
Honey bee	Rat	Trout	Trout
Bumble bee	Rabbit	Frog	Rabbit adductor longus
African dung beetle	Mouse	Toad	Chicken pectoral muscle
Rosechafer	Sheep	Radular retractor of:	
Summer chafer	Chicken	Limpet	
Cockroach		Periwinkle	
Locust (African and desert)		Red muscles of:	
Blowfly		Dogfish	
Fleshfly		Trout	
Housefly		Pectoral muscles of:	
Tsetse fly		Pheasant	
		Pigeon	
		Rat soleus muscle	
		Rat diaphragm	
		Human thigh muscle	
		Rabbit semitendinosus	
		Toad sartorius	

contains vertebrate muscles which are considered predominantly 'red'. These may be ranked according to the maximum flux through the TCA cycle as follows: pigeon pectoral > rat soleus and rat diaphragm > dogfish red muscle > trout red muscle > pheasant pectoral muscle > toad sartorious. (The finding regarding pheasant pectoral muscle is anomalous since it is considered that this muscle is largely anaerobic: Crabtree & Newsholme, 1972; Newsholme & Start, 1973.)

It is interesting that the sartorius muscle of the African clawed toad (*Xenopus laevis*) comes into group 3 (see Table 3). The name of this animal is a misnomer, as it is classified as a frog (Frazer, 1973), but the animal is entirely aquatic and is a powerful swimmer, and therefore not surprisingly the activities of oxoglutarate in the sartorius and gastrocnemius are greater than those of the muscles in the common British frog, *Rana temporaria*. It is suggested that *Xenopus* muscles require a greater aerobic capacity to support continuous swimming activity than the skeletal muscles of *Rana*, which may rely more on anerobic energy production to support the occasional bursts of contraction required in jumping.

Group 4 contains predominantly 'white' vertebrate muscles. Evidence for the anaerobic nature of white muscles has been discussed by Needham (1971). The availability of data on the maximum activity of oxoglutarate dehydrogenase provides the first quantitative confirmation of the low aerobic capacity of vertebrate white muscles. A comparison of the maximum rate of ATP production from aerobic and anaerobic pathways can now be made with knowledge of the activities of 6-phosphofructokinase as an index of anaerobic glycolysis (Newsholme & Crabtree, 1972*a*; Newsholme *et al.*, 1978) and oxoglutarate dehydrogenase as an index of the TCA cycle.

Comparison of aerobic fuels for different animals

If the maximum activity of oxoglutarate dehydrogenase provides a quantitative indication of the maximum flux through the TCA cycle, it is possible to compare these data with the flux into the cycle from carbohydrate and lipid fuels to suggest which could be the more important under aerobic conditions. The flux from glucose could be indicated by the maximum activities of hexokinase (see Crabtree & Newsholme, 1972*a*; Newsholme *et al.*, 1978; Surholt & Newsholme, 1981) and the flux from lipids (either triglyceride, diglyceride or fatty acids) by the maximum activities of triglyceride lipase, diglyceride lipase or carnitine palmitoyltransferase (Crabtree & Newsholme, 1972*b*). These comparisons are given for selected animals or groups of animals below.

Insects

The high concentrations of fuels in the haemolymph of insects may reflect the difficulty of distributing fuels to the muscles; the 'open blood system' of invertebrates with the haemolymph bathing all the organs of the body is not likely to be as efficient for the delivery of fuels to the individual muscles as the 'closed blood system' of the vertebrates, so that high concentrations of fuels are necessary to facilitate diffusion to and uptake into the muscle fibres. On the other hand, the efficiency of the tracheolar system in distributing oxygen to the flight muscles is extremely high so that adequate oxygen is supplied to meet the demands of flight so that they have a high capacity of the TCA cycle and a low capacity to produce lactate.

The availability of oxoglutarate dehydrogenase activities for insect flight muscle provides, for the first time, a quantitative indication of the importance of the TCA cycle. In addition, it allows a comparison between hexokinase activity and the capacity of the TCA cycle, which indicates the probable importance of glucose oxidation for energy production for insect flight. In most insects investigated, utilisation of glucose (which may be obtained from trehalose in the haemolymph) can provide sufficient acetyl-CoA to satisfy much of the capacity of the TCA cycle. However, although glucose and/or trehalose concentrations in the haemolymph are high, this store of carbohydrate would not be sufficient to support sustained flight for several hours.

The advantage of fat as a storage fuel, especially in flying animals, has been stressed (Newsholme & Start, 1973). However, the transport of such a fuel in the haemolymph is a problem, especially in insects since very high concentrations of fuels are necessary. This fuel cannot be free fatty acids, since these are toxic unless bound to protein and high concentrations of protein may produce an adverse osmotic balance. Hence fat is transported as diglyceride.

Why proline should be used as a fuel by some insects is unclear. There is increasing evidence that it is used in roaches and beetles. The pathway for utilisation of this fuel produces alanine as the end-product; there may be little or no osmotic change involved in converting proline to alanine and this may be beneficial in some insects.

Fish

Fish possess two discrete types of muscle: white and red. The white muscle forms the main muscle mass and is used in violent bursts of swimming; the lateral strip of red muscle, forming only 8% of the total weight, is used in normal cruise swimming (Bone, 1966). This lateral muscle contains a large number of mitochondria and lipid droplets and high activities of cytochrome oxidase and lipolytic enzymes. It has always been assumed that such characteristics correlate with a high aerobic capacity, but it has not been possible to test this suggestion quantitatively until it was established that activities of oxoglutarate dehydrogenase provide such information.

In the present study, two fish have been investigated: one a teleost, the trout, which is an active swimming fish, and the other an elasmobranch, the dogfish, which may spend much time in the substratum (Bone, 1966) and does not possess the streamlining (as an adaptation to fast swimming) that the trout does. The storage of lipid is also different in these two fish. Elasmobranchs store most of their triglyceride in the liver, while teleosts store most of theirs in an abdominal adipose tissue and/or in adipocytes which are distributed throughout the muscle fibres (Love, 1970).

Comparison of the activities of other fuel-utilising enzymes with the maximum calculated flux through the TCA cycle shows that carbohydrate oxidation could support maximum energy requirements in both types of muscle in both fish. In the dogfish, the activities of triglyceride lipase and carnitine palmitoyltransferase are very low and comparison with those of oxoglutarate dehydrogenase suggests that fatty acids can support less than 4% of the flux through the TCA cycle; this in turn suggests that fatty acids are an unimportant fuel for swimming in this fish. Ketone bodies, however, are important in the dogfish, in common with other elasmobranchs (Zammit & Newsholme, 1979). In the trout, comparison of the activities of carnitine palmitoyltransferase with oxoglutarate dehydrogenase suggests that fatty acid oxidation could support 44% of the maximum flux through the TCA cycle in red muscle. However, in the red muscle of the mackerel, activities of the transferase suggest that flux through the TCA cycle (2.4 μmol min^{-1} g^{-1}) could be supported entirely from oxidation of fatty acids (Zammit & Newsholme, 1979).

Birds

In sustained flight of migratory birds, the respiratory exchange ratio ranges from 0.7 to 0.8, suggesting that fat is a major fuel. Indeed, fat amounting to 20–50% of the body weight may be used up during migration (Tucker, 1972). Although migratory birds have not been investigated in this study the pigeon is known to fly long distances and its pectoral muscle contains large numbers of mitochondria and lipid droplets suggesting that fat is an important fuel (see Newsholme & Start, 1973). The activity of oxoglutarate dehydrogenase provides a quantitative indication of the importance of TCA cycle in the pectoral muscle of the pigeon and consequently of migratory birds.

Comparisons of enzyme activities indicates that in pigeon pectoral muscle endogenous triglyceride could provide 53% and fatty acid oxidation 100%

of the maximum capacity of the TCA cycle. This is particularly important, since the amount of carbohydrate stored in such birds must be very limited; weight versus lift is crucial in the power output of flying animals and fat is a much better storage form of fuel because it can be stored in the absence of water (see Newsholme & Start, 1973).

The two other birds investigated are the chicken (a domesticated galliform) and the pheasant (also a member of the Galliformes). These game birds are mainly ground-living and possess only a limited capacity for flight, which is restricted to flapping into the air and gliding down to cover (Welty, 1955). The activity of oxoglutarate dehydrogenase indicates that the maximum capacity of the TCA cycle in the pectoral muscle of the chicken is very low (< 0.1 μmol min^{-1} g^{-1}) and this is the first quantitative indication of the lack of aerobic capacity of this muscle. Somewhat surprisingly, the maximum flux through the TCA cycle is higher in the pheasant pectoral muscle (1.5 μmol min^{-1} g^{-1}). The difference between the chicken and pheasant pectoral muscles may be due to the domestication of the chicken.

Mammals

Mammalian skeletal muscle usually contains a mixture of red and white fibres. The muscles investigated in the present study include the rat soleus (predominantly red), the rabbit semitendinosus (predominantly red) and the rabbit adductor longus (predominantly white). The activity of oxoglutarate dehydrogenase and hence the maximum capacity of the TCA cycle in the rabbit adductor longus muscle is very low, which supports previous assumptions that mammalian white skeletal muscle is mainly anaerobic (see Needham, 1971). The activities of triglyceride lipase and carnitine palmitoyltransferase are available for the rabbit muscles, and for the rat quadriceps. The deeper fibres of the rat quadriceps are more red than the outer fibres and may have a similar metabolism to the rat soleus (Rennie & Holloszy, 1977), so that the activities of hexokinase, lipase and transferase activities can be compared with those of oxoglutarate in the soleus muscle. From this comparison it appears that in the soleus muscle, muscle triglyceride could support 8% of the maximum TCA cycle flux whereas fatty acids could support 100%. Similarly, in the rabbit semitendinosus, triglyceride and fatty acid oxidation could contribute 70% and 100% repectively to the maximum flux through the TCA cycle.

References

Alp, I. R., Newsholme, E. A. & Zammit, V. A. (1976). Activities of citrate synthase, NAD$^+$-linked and NADP$^+$-linked isocitrate dehydrogenase in muscle from vertebrates and invertebrates. *Biochemical Journal*, **154**, 689–700.

Beenakkers, A. M. Th. (1969). Carbohydrate and fat as a fuel for insect flight. A comparative study. *Journal of Insect Physiology*, **15**, 353–61.

Bone, Q. (1966). The function of the two types of myotomal muscle fibre in elasmobranch fish. *Journal of the Marine Biological Association of the United Kingdom*, **46**, 321–49.

Cooney, G., Taegmeyer, H. & Newsholme, E. A. (1981). Tricarboxylic acid cycle flux and enzyme activities in the isolated working heart. *Biochemical Journal*, **200**, 701–7.

Crabtree, B. & Newsholme, E. A. (1972a). The activities of phosphorylase hexokinase, phosphofructokinase, lactate dehydrogenase and glycerol-3-phosphate dehydrogenase in muscles from vertebrates and invertebrates. *Biochemical Journal*, **126**, 49–58.

Crabtree, B. & Newsholme, E. A. (1972b). The activities of lipases and carnitine palmitoyltransferase in muscles from vertebrates and invertebrates. *Biochemical Journal*, **130**, 697–705.

Crabtree, B. & Newsholme, E. A. (1975). Fuel supply in insects. In *Insect Muscle*, ed. P. N. R. Usherwood, pp. 405–94. New York & London: Academic Press.

Crabtree, B. & Taylor, D. J. (1979). Thermodynamics and metabolism. In *Biochemical Thermodynamics*, ed. M. N. Jones, pp. 333–78. Amsterdam: Elsevier/North-Holland.

Frazer, J. F. P. (1973). *Amphibians*. London: Wykeham Publications.

Hazel, J. A. & Prosser, C. L. (1974). Molecular mechanism of temperature compensations in poikilotherms. *Physiological Reviews*, **54**, 620–77.

Jones & Sidell, B. D. (1982). Metabolic responses of striped bass (*Morone saxatilis*) to temperature acclimation. II. Alterations in metabolic carbon sources and distributions of fibre types in locomotory muscle. *Journal of Experimental Zoology*, **219**, 163–71.

Hill, A. V. (1950). The dimensions of animals and their muscular dynamics. *Science Progress*, **38**, 209–30.

Love, R. M. (1970). *The Chemical Biology of Fishes*. New York & London: Academic Press.

Needham, D. M. (1971). *Machina Carnis*. Cambridge University Press.

Newsholme, E. A. & Crabtree, B. (1976). Substrate cycles in metabolic regulation and heat generation. *Biochemical Society Symposia*, **41**, 61–110.

Newsholme, E. A. & Crabtree, B. (1979). Theoretical principles in the approaches to control of metabolic pathways and their application to glycolysis in muscle. *Journal of Molecular and Cellular Cardiology*, **11**, 839–56.

Newsholme, E. A. & Crabtree, B. (1981). Control of flux through metabolic pathways. In *Short Term Control in the Liver*, ed. L. Hue & G. van der Werve, pp. 3–18. Amsterdam: Elsevier/North-Holland.

Newsholme, E. A., Crabtree, B. & Zammit, V. A. (1980). Use of enzyme activities as indices of maximum rates of fuel utilization. *Ciba Foundation Symposia*, **73**, 245–58.

Newsholme, E. A. & Start, C. (1973). *Regulation in Metabolism*. New York: Wiley.

Newsholme, E. A., Zammit, V. A. & Crabtree, B. (1978). Carbohydrate fuels for muscle. *Biochemical Society Transactions*, **6**, 512–20.

Owen, G. (1966). Feeding. In *Physiology of Mollusca*, vol. 2, ed. K. M. Wilbur & C. M. Yonge, pp. 1–51. New York & London: Academic Press.

Paul, J. (1979). Comparative studies on the citric acid cycle. DPhil thesis, University of Oxford.

Pette, D. (1966). Mitochondrial enzyme activities. In *Regulation of Metabolic Processes in Mitochondria*, eds. J. M. Tager, S. Papa, E. Quagliariello & E. C. Slater, pp. 28–50. Amsterdam: Elsevier.

Rennie, M. J. & Holloszy, J. O. (1977). Inhibition of glucose uptake and glycogenolysis by availability of oleate in well oxygenated perfused skeletal muscle. *Biochemical Journal*, **168**, 161–70.

Rowan, A. N. & Newsholme, E. A. (1979). Changes in contents of adenine nucleotides and intermediates of glycolysis and citric acid cycle in flight muscle of locust upon flight and their relationship to the control of the cycle. *Biochemical Journal*, **178**, 209–16.

Shaklee, J. B., Christiansen, J. C., Sidell, B. D., Prosser, C. L. & Whitt, G. S. (1977). Molecular aspects of temperature acclimation in fish: the contributions of changes in enzyme activities and isozyme patterns to metabolic reorganisation in the green sunfish. *Journal of Experimental Zoology*, **210**, 1–20.

Stone, B. B. & Sidell, B. D. (1982). Metabolic responses of striped bass (*Morone saxatilis*) to temperature acclimation. I. Alterations in carbon source for hepatic energy metabolism. *Journal of Experimental Zoology*, **218**, 371–9.

Tucker, V. A. (1976). Respiration during flight in birds. *Respiratory Physiology*, **14**, 75–82.

Weiss-Fogh, T. (1961). Power in flapping flight. In *Cell and the Organism*, ed. J. R. Ramsey & V. B. Wigglesworth, pp. 283–300. Cambridge University Press.

Welty, C. (1955). Birds as flying machines. *Scientific American*, **192**, 88–96.

Zammit, V. A. & Newsholme, E. A. (1979). Activities of enzymes of fat and ketone body metabolism and effects of starvation on blood concentrations of glucose and fat fuels in teleost and elasmobranch fish. *Biochemical Journal*, **184**, 313–22.

BRUCE D. SIDELL

Cellular acclimatisation to environmental change by quantitative alterations in enzymes and organelles

Metabolism and the ability to respond to changes in the environment are characteristics shared by all living cells. The relationship between these features is most evident when one considers that alterations in metabolism, i.e. chemical reactions mediated in living systems, are often among the most pronounced responses of cells to environmental stimuli, such as temperature, pressure, oxygen availability, nutrition, light or even pathogens. Cellular metabolism may be readjusted to meet environmental challenges by a variety of alternative mechanisms. The nature of response often depends upon how rapidly the stimulus is presented and/or how rapidly the organism must react to prevent metabolic derangement or failure.

Within the cell an intricate chemical network of feedback loops permits swift activation or inhibition of key regulatory enzymes to co-ordinate response to acute metabolic demands. For example, glycolytic flux in skeletal muscle may increase almost instantaneously by up to 100-fold at the onset of activity (Newsholme & Start, 1973). At the opposite temporal extreme of adaptation, evolutionary time, it is apparent that protein structures have been selected and genotypes fixed for most efficient function within the limits of the species' normal environmental regime. Characteristics of enzyme function, such as ligand binding affinities, have been strongly conserved in homologous enzymes of species from both tropical and polar environments (Somero, 1978). For example, muscle-type lactate dehydrogenases (M_4-LDHs) from fishes of differing thermal environments show marked similarity for both apparent Michaelis constant (K_m) of pyruvate and catalytic constant (k_{cat}) when measured at normal cell temperature and pH (Yancey & Somero, 1978).

At an intermediate point in this temporal spectrum cellular function within individual organisms may acclimatise to nutritional, thermal or other stresses over a period of days, weeks or months. Unlike the adaptive 'strategies' employed on an evolutionary time-scale, in which different protein structures have evolved in different environments, the mechanisms underlying cellular acclimatisation of these organisms must lie within the genotypic limits of the individual. Alterations in intracellular concentrations of enzymes have now been shown to affect many metabolic readjustments during acclimatisation. This article explores the role played by variations in the quantity of enzymes and organelles during cellular acclimatisation to environmental change.

Adaptive value of adjusting enzyme concentration

A wide variety of physiological or environmental stimuli can elicit changes in cellular protein concentration in many organisms (a few examples are given in Table 1). Given the possible scope of response which can be realised by either activating or inhibiting existing enzyme pools within the cell, it is reasonable to question the adaptive advantage of the more 'brute force' mechanisms of altering enzyme concentration. Arguments considering metabolic regulation, energetics and physical constraints of the cell may be marshalled to support the value of readjusting enzyme level in response to long-term metabolic demand.

Table 1. *Examples of cellular proteins which change in concentration after environmental or physiological stimuli*

Protein	Animal/tissue	Stimulus	Reference
Arginase	Rat liver	Dietary protein level	a
Xanthine oxidase	Rat liver	Dietary protein level	a
NADPH–cytochrome c reductase	Rat liver	Phenobarbitol	a
Ferritin	Rat liver	Dietary iron	b
Carbamyl phosphate synthetase	Frog liver	Thyroxine	a
Malic enzyme	Chick liver	Feeding/fasting	c
Cytochrome oxidase	Fish muscle	Temperature	d
Cytochrome c	Rat muscle	Exercise	e
	Fish muscle	Temperature	f
Myoglobin	Fish muscle	Temperature	g
Cytochrome P-450	Armyworms	Petroleum hydrocarbons	h
Haemoglobin	*Daphnia*	Oxygen	i
Chlorophyll	*Euglena*	Light	j

[a] Schimke & Doyle (1970); [b] Goldberg & Dice (1974); [c] Goodridge (1975); [d] Wilson (1973); [e] Terjung *et al.* (1975); [f] Sidell (1977); [g] Sidell (1980); [h] Wilkinson (1979); [i] Fox (1955); [j] Zeldin *et al.* (1973).

Maintenance of metabolic regulation

Physiological concentrations of cellular metabolites frequently are matched closely to the K_m values of enzymes for which they serve as substrates. Indeed, activation or inhibition of many regulatory enzymes is affected by ligand-induced modulation of their K_m-values while substrate concentration remains relatively stable (Hochachka & Somero, 1973). Matching of K_m and substrate concentration range is thus critical to maintenance of regulatory sensitivity of most allosteric enzymes. Activation of a rate-limiting enzyme by this mechanism in response to long-term metabolic demand would result in an increase in rate (closer to maximum velocity, V_{max}), but the ability to maintain metabolic control at this point would be greatly diminished in the process. An increase in cellular concentration of the same enzyme under these circumstances permits similar compensation in the rate of product formation while conserving the crucial relationship between K_m and substrate concentration with its inherent regulatory sensitivity (Fig. 1).

Energy economy of the cell

Because of the molecular size and complexity of enzymes, each enzyme molecule within the cell represents a considerable investment of

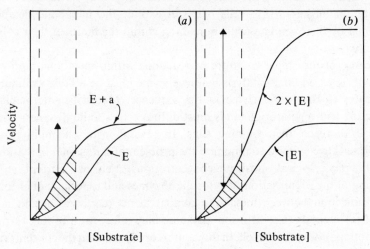

Fig. 1. Two mechanisms of enhancing cellular enzyme activity. (*a*) Allosteric activation by a regulatory ligand (a = presence of activator). (*b*) Elevation in cellular enzyme concentration. The cross-hatched area of each curve indicates the rate increase achieved within the range of physiological substrate concentration (delineated by broken lines). Note that the available scope for further activation of the enzyme (indicated by arrows) is much greater at elevated enzyme concentration (*b*).

energy. At the same time, high activity of some metabolic pathways is only necessary under restricted physiological conditions. For example, maintenance of a large pool of ornithine–urea cycle enzymes in the mammalian liver is vital for nitrogen metabolism of animals fed a high-protein diet, but represents an under-utilised resource during conditions of low protein intake. By lowering enzyme levels through degradation under the latter circumstances, constituent amino acids are made available for either energy metabolism or more critical needs of protein synthesis. Responding to metabolic demand by altering enzyme concentration may thus be viewed as a favourable energetic 'investment strategy' for the cell.

Physical considerations

When considered as an integrated physicochemical system, it becomes apparent that the living cell is also subject to certain physical constraints. Although water accounts for 65–80% of the mass of most cells, its solvent capacity is not limitless. The average mammalian somatic cell has been estimated to have the genetic capacity to code for approximately 10^6 different types of protein molecules. The simultaneous presence of each of these proteins at functionally meaningful levels would not only pose severe energetic problems, but would also greatly outstrip the capacity of the cell to hydrate them. Not unexpectedly, we find that sufficient messenger RNA molecules to code for only of the order of 10^4 different proteins are present in these cells at any time (Lewin, 1980). A strategy of maintaining 'standing armies' of each of the thousands of known enzymes of the cell, even if energetically feasible, is precluded therefore strictly on the basis of available cellular water. By increasing the concentration of specific enzymes when they are needed and reducing the concentration of others when demand for their activity diminishes, living cells can still exploit the metabolic flexibility provided by their genome while remaining within the limits of their solvent capacity.

Because of the stepwise nature of reactions within metabolic pathways, products released after catalysis by one enzyme must be readily available as substrates to the next enzyme in the sequence. Metabolic intermediates, co-factors and regulatory ligands must diffuse across finite distances within the cell to reach their binding sites. In several instances multi-enzyme complexes have evolved to minimise the problem of diffusional limitation at realistic physiological substrate concentrations. Yet, our current understanding of the organisation of cytosolic enzymes and especially of chemical communication between metabolic compartments (e.g. cytosol and mitochondria) suggests that diffusion distances may affect rates of flux through metabolic pathways of the cell. In this context an elevation in the concentration of enzymes or organelles in response to long-term metabolic stimuli or physical factors that affect rates of diffusion (e.g. temperature) may significantly reduce mean path-lengths of diffusion for cellular metabolites.

Each of the above considerations has focused on the adaptive advantages of changing enzyme concentration, but certain limitations are also implicit in this mode of response. Foremost is that hours, days or even weeks usually must elapse before new steady-state concentrations of enzymes can be achieved. In light of the argument on energetics presented above, it is also clear that significant cellular energy reserves are tied up, at least temporarily, in high concentrations of enzyme. Finally, the aspect of solvent capacity is also a double-edged sword which intrinsically sets limits on the concentrations and number of enzymes whose level may be adjusted upward. Like the structure/function 'compromises' for protein evolution recently reviewed by Somero (1978), selective pressures have balanced these advantages and limitations to dictate the nature of cellular acclimatisation to environmental change.

Mechanisms of altering enzyme levels

At least two major avenues are available for changing the concentration of active enzymes within the cell: (1) readjustment of the current rates of synthesis and breakdown of the enzyme, i.e. the rate of turnover of the enzyme, and (2) enzymatic interconversion between existing pools of active and inactive forms, frequently by reversible phosphorylation of the enzyme. Several examples of the latter mechanism may be found in any biochemistry text and will not be considered further here.

Far from being a static complement, the levels of essentially all protein constituents of animal cells are determined by the balancing of the specific protein's continuous rates of synthesis and of degradation (Goldberg & Dice, 1974; Dean, 1978). Any mathematical description of the system must include terms for both synthesis and degradation. An accepted expression used to describe changes in enzyme levels (Schimke, 1970) is:

$$dE/dt = S - k_d[E] \qquad (1)$$

where [E] represents the concentration of the enzyme (units·mass^{-1}), S is the rate of synthesis (units·time^{-1}·mass^{-1}) and k_d is the first-order rate constant for degradation (time^{-1}). Although the rate of synthesis of an enzyme is undoubtedly a first-order process with respect to such variables as the availability of transfer RNA, amino acids, messenger RNA and several other translation factors, measurements of these variables *in vivo* are generally not available in studies of protein turnover. An overall zero-order term for the rate of enzyme synthesis has proved adequate in such treatments. In the steady-state condition (with respect to enzyme concentration):

$$dE/dt = 0 = S - k_d[E]$$

Therefore

$$S = k_d[E] \quad \text{or} \quad [E] = S/k_d \tag{2}$$

Thus, even in the steady-state condition, the amount of enzyme present is the result of both the rate of synthesis and the rate of degradation of the specific enzyme. By altering either of these rates, or both, the concentration of an enzyme within the cell may be increased or decreased.

Technical considerations
Quantifying the concentration of enzymes and organelles

Direct measurement of the quantity of a specific enzyme in tissue extracts is possible either by immunochemical methods or by spectrophotometric techniques in the case of proteins with unique spectral characteristics (e.g. haemoproteins). Production of antibodies against specific enzymes is a laborious task and subsequent quantification of tissue levels of enzyme requires careful control of precipitation conditions. Direct spectrophotometric methods are obviously restricted to a limited number of cellular proteins such as mitochondrial cytochromes or respiratory pigments. Consequently, most studies aimed at assessing cellular concentrations of enzymes have relied upon indirect determinations by measurement of differences in activity between physiological treatments. Maximum tissue activities of enzymes, measured under optimal assay conditions and at saturating substrate concentration, are assumed to reflect accurately the number of enzyme molecules present. This approach appears valid in many cases. However, changes in phospholipids associated with membrane-bound enzymes may alter their specific activity and obscure shifts in enzyme content (reviewed by Hazel & Prosser, 1974). Results obtained by enzyme activity measurement are thus inferential and caution must be exercised in interpreting such data.

Our ability to assess accurately changes in density and morphometry of the subcellular organelles in which many enzymes are situated is less equivocal. Modern computer technologies and stereological techniques have been successfully applied to analysis of electron micrographs of biological specimens (Meek & Elder, 1977). Through such studies we have gained important insight into the ultrastructural 'plasticity' that cells can show in response to environmental change.

Analysis of protein turnover

During the last 15 years great strides have been made towards overcoming the many experimental pitfalls encountered in measuring the turnover rates of specific enzymes. The majority of studies utilise one of two experimental approaches: (1) following the decay of specific radioactivity of the enzyme of interest after administration of a radioactively labelled precursor, or (2) measurement of the rate of change in concentration or activity of enzyme during the transition between physiological steady states. Each of these methodologies has been admirably reviewed by Waterlow, Garlick & Millward (1978), Goldberg & Dice (1974) and Schimke & Doyle (1970).

With this theoretical and technical background established, we may now turn to concrete examples of cellular adaptation based upon quantitative changes in the content of enzymes or organelles. The remainder of this article focuses on cellular and biochemical adaptations of fishes to changes in environmental temperature, a system that is particularly well suited to studies of metabolic regulation.

Subcellular concentration changes during thermal adaptation

It is often difficult to identify the specific level at which environmental variables affect cellular function in higher animals. For example, homeostatic responses by the cardiovascular and respiratory systems can ensure little variation in oxygen supply to core tissues, even when the availability of oxygen in the environment varies over a wide range. The body temperature of most fishes and other aquatic poikilotherms, however, fluctuates directly with their thermal surroundings. For these animals temperature is one of the few physical variables which places the environment–organism interface at the molecular level. Despite the profound effect of temperature on biochemical reaction rate, groups of fishes have evolved to exploit a remarkable range of thermal niches around the globe, and individuals of eurythermal species can encounter large seasonal variations in water temperature at high-temperate latitudes. The latter group is of considerable interest with respect to cellular acclimatisation. Across wide seasonal temperature ranges, many of these species display a strong conservation of biological activity, both in terms of metabolic energy 'production' and energy 'consumption' (biological work).

Metabolic temperature compensation

The immediate response in the metabolic rate of fishes to a shift in environmental temperature is predictable on the basis of physical chemistry; increases in water temperature accelerate metabolism while decreases in water temperature decelerate metabolism. Yet, it has long been known from laboratory experiments on thermal acclimation that many fish species show the capacity to compensate their metabolism over a period of days, weeks or months to overcome the initial kinetic effect of temperature change. During acclimation to cold, oxygen consumption of whole animals can gradually increase from initially depressed rates to approach those of animals acclimated to much warmer temperatures. Similar responses to thermal acclimation are often measured at the level of cellular respiration although the pattern and extent of metabolic compensation may differ considerably between isolated tissues from the same fish (Hazel & Prosser, 1974). The magnitude of these shifts is often impressive. When measured at their respective temperatures of acclimation, some isolated tissues from cold-acclimated animals may actually show a greater metabolic rate than those from warm-acclimated fish, as we have recently found with skeletal muscle of striped bass (*Morone saxatilis*) (Fig. 2; Jones & Sidell, 1982).

The last two decades have witnessed vigorous research efforts aimed at understanding the molecular mechanisms that underlie metabolic compensation to temperature (reviewed by Hazel & Prosser, 1974; Somero, 1978).

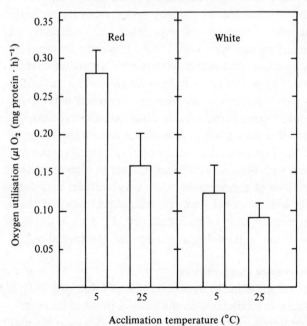

Fig. 2. Rate of oxygen utilisation by red and white muscle from thermally acclimated striped bass. Measurements were made *in vitro* at the acclimation temperature of each animal. Mean ± S.E.M. for five fish at each temperature. (From Jones & Sidell, 1982.)

Biochemical studies of fish tissues have revealed significant acclimation effects on the maximal activities of some enzymes from pathways of energy metabolism while the activities of others remain unaffected. Increases in the catalytic capacity of enzymes during cold acclimation (most notably those from aerobic pathways) have been interpreted as an adaptive response to at least partially offset acute temperature effects on reaction rate. Like the case for oxygen consumption, enzymic responses to thermal acclimation in fish can be highly tissue-specific (Shaklee et al., 1977), presumably reflecting a major reorganisation of the animal's metabolism. The enzymes of skeletal muscle have been especially widely studied and provide an excellent system to examine more closely.

Table 2. *Patterns of activity or concentration response to temperature acclimation for enzymes of energy metabolism in fish muscle*

Enzyme/protein	Species	5 °C-acclimated/ 25 °C-acclimated	Reference
Cytochrome oxidase	Goldfish	1.66	a
	Green sunfish	1.94	b
	Striped bass (red fibres)	1.97	c
Cytochrome c	Green sunfish	1.51	b, d
Succinic dehydrogenase	Goldfish	1.84	e
	Green sunfish	1.35	b
Citrate synthase	Striped bass (red fibres)	1.57	c
Glucosephosphate isomerase	Green sunfish	0.83	b
Phosphofructokinase	Striped bass (red fibres)	1.03	c
Glyceraldehyde-3-phosphate dehydrogenase	Green sunfish	0.85	b
Pyruvate kinase	Green sunfish	0.74	b
	Striped bass (red fibres)	0.66	c
Lactate dehydrogenase	Goldfish	0.74	e
	Green sunfish	0.90	b
	Striped bass (red fibres)	0.92	c

[a] Wilson (1973); [b] Shaklee et al. (1977); [c] Jones & Sidell (1982); [d] Sidell (1977); [e] Sidell (1980).

Enzyme concentration and turnover

On the basis of arguments already presented, we can predict increases in the concentration of those enzymes from muscle that show an elevated activity in tissues of cold-acclimated animals (Table 2). To date, however, only two studies have unequivocally confirmed this prediction. By using immunochemical techniques, the content of cytochrome oxidase was shown to be 66% greater in skeletal muscle of goldfish (*Carassius auratus*) acclimated to 5 °C than in those acclimated to 25 °C (Wilson, 1973). Parallel measurements of the maximal catalytic activity of this mitochondrial enzyme showed a 45% increase. Concentration of another member of the electron transport chain, cytochrome c, also increases in fish muscle during cold acclimation. Spectrophotometric determinations show that the level of cytochrome c in skeletal muscle of green sunfish (*Lepomis cyanellus*) rises with decreasing acclimation temperature over a range of 25 °C to 5 °C (Fig. 3; Sidell, 1977). The environmentally induced change in concentration of this protein results from different thermal sensitivities of the rates of synthesis and degradation.

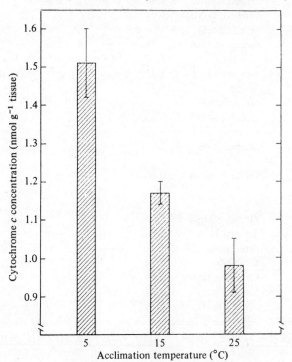

Fig. 3. Effect of acclimation temperature on the concentration of cytochrome c in skeletal muscle of green sunfish (mean±S.E.M.). 5 °C, $n = 5$; 15 °C, $n = 6$; 25 °C, $n = 5$. (Data are from Sidell, 1977.)

Because cytochrome c can be quantitatively extracted from muscle and selectively trapped by ion-exchange chromatography, it has proved especially well suited to investigations of protein turnover in response to environmental temperature. Data obtained by analysis of the rate of change in concentration of cytochrome c during acclimation of green sunfish to 5 °C or 25 °C (Fig. 4) and by measurement of the loss in specific radioactivity of the cytochrome, were in excellent agreement (Sidell, 1977). The radioisotope used was [δ-^{14}C]aminolaevulinic acid, which is a non-reutilisable precursor of haem. Exposure of fish to cold temperature resulted in a rapid decrease in the rate of synthesis of skeletal muscle cytochrome c, but the rate constant for degradation of this protein was reduced even more. Because the effect of temperature on these rates is disproportional, the concentration of cytochrome c increases during cold acclimation, or decreases during warm acclimation until the relationship described in equation (2) is balanced at the new steady state. These findings on the turnover of cytochrome c have two interesting

Fig. 4. Time-course of change in cytochrome c concentration of green sunfish skeletal muscle during acclimation between 5 °C and 25 °C. Mean ± S.E.M.; $n = 4$ at each point. S, zero-order rate constant for synthesis; k_d, first-order rate constant for degradation; $t_{\frac{1}{2}}$, time taken for decay to 50% of control value. (Adapted from Sidell, 1977.)

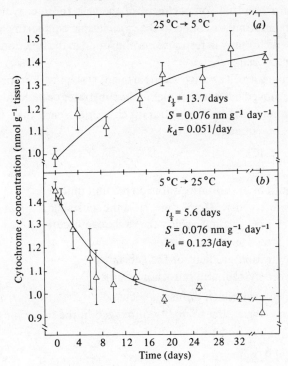

implications. First, the observed differential changes in the rates of synthesis and degradation enable the organism to achieve an increase in the content of an enzyme while actually reducing the rate of enzyme synthesis. An adaptive increase in enzyme concentration at cold temperature is thus realised with minimum energetic demand for protein synthesis. Secondly, the data suggest that the change in concentration of cytochrome c may result from a passive effect of temperature on the rate constants for synthesis and degradation, although an active control at higher levels of organisation (e.g. hormonal or nervous integration) is not ruled out.

Organelle concentration

The relationships between proteins that change in concentration during temperature adaptation seem more than coincidental. Both cytochrome c and cytochrome oxidase are components of the mitochondrial inner membrane. In addition, the half-life of the total pool of mitochondrial haem proteins in green sunfish muscle is identical to that found for cytochrome c (Sidell, 1977), suggestng that the cytochrome c turnover may be representative of that of other members of the respiratory chain. Indeed, closer inspection of the patterns of enzymic response to temperature in fish muscle (Table 2) reveals that each enzyme showing marked compensation to temperature is associated with the mitochondrial compartment of the cell; those enzymes showing a lack of compensation are found in the cytoplasmic compartment. A possible basis for this dichotomy in response may be found in the functional morphology of the muscle cell.

If one considers the muscle cell as a two-compartment system with respect to metabolic processes, movement of either metabolic substrates or molecular oxygen from the extramitochondrial to intramitochondrial space may be described by the diffusion equation:

$$\mathrm{d}n/\mathrm{d}t = -D \cdot A \left(\frac{\Delta C}{X}\right)$$

where $\mathrm{d}n/\mathrm{d}t$ = the amount of substance diffusion per unit time;
D = diffusion coefficient (cm² time⁻¹) for the substance;
A = surface area through which the exchange occurs (mitochondrial surface area);
C = concentration gradient of the substance (between cytosol and mitochondrion);
X = distance over which diffusion occurs.

In the case of molecular oxygen, the solubility of oxygen in the tissue must be included, and the equation becomes:

$$\dot{V}_{O_2} = -K_{O_2} \cdot A \left(\frac{\Delta P_{O_2}}{X}\right)$$

where A and X are the same as above;

$\dot{V}_{O_2}$ = the rate of oxygen delivery;

K_{O_2} = Krogh's constant of diffusion for oxygen through muscle tissue (incorporating both D and oxygen solubility);

ΔP_{O_2} = the partial pressure gradient for oxygen (in this case, $P_{O_2 \text{ (blood)}} - P_{O_2 \text{ (mitochondrion)}}$).

Empirical evidence is available to support the validity of this treatment. The movement through muscle cells of oxygen (Mahler, 1978) and of several ions and non-charged solutes (Kushmerick & Podolsky, 1969; Caille & Hinke, 1974) has been shown to conform to the one-dimensional diffusion equation.

The lowered environmental/cell temperature encountered during cold acclimation will decrease D for metabolites and oxygen because of both the direct effect of temperature on molecular kinetic energy and the resulting increase in solvent viscosity of the cell. Indeed, the diffusion coefficient for oxygen in frog skeletal muscle has been shown to decrease more than 40% between the temperatures of 23 °C and 0 °C (Mahler, 1978). Since respiration and its tight regulation are dependent upon continuous supply of adenylates, metabolites and oxygen to the mitochondrial compartment, how can the high resting rates of oxygen consumption found in muscle of cold-acclimated animals (Fig. 2) be reconciled with the effect of temperature on diffusion of these cellular commodities?

In the case of oxygen it appears that the increase in oxygen capacitance of the tissue at cold temperatures and an increase in cellular myoglobin concentration during cold acclimation (Sidell, 1980) may combine to offset thermal effects on diffusivity. The situation for adenylates and metabolites is less clear-cut, because temperature has no effect on their solubility and concentrations apparently remain stable. The recent finding of an increase in the mitochondrial population of muscle fibres in cold-acclimated fishes may help explain this apparent paradox.

Acclimation from 28 °C to 2 °C results in a significant increase in mitochondrial density in all muscle fibre types in crucian carp (*Carassius carassius*) (Johnston & Maitland, 1980). A similar cold-induced elevation in the fraction of cell volume occupied by mitochondria and in the density of mitochondrial profiles has been observed in red and white muscle of goldfish after acclimation from 25 °C to 5 °C (Table 3; Fig. 5).

An increase in the mitochondrial proportion of cell volume at cold temperatures will elevate the A term of the diffusion equation by increasing the area of mitochondrial surface per cell, and may also decrease the X term, the mean intracellular diffusion path-length. Both of these changes compensate for the depression of D for adenylates and metabolites at cold temperature. Importantly, the lack of significant difference in the surface-to-volume ratio and the concomitant increase in mitochondrial fraction of cell volume during cold acclimation (Table 3) imply a proliferation of the mitochondrial population within the cell, rather than simple enlargement of those organelles already present. On the basis of these new stereological data, we may conclude that the cold-induced rise in concentrations of mitochondrial enzymes of fish muscle may *not* be solely to overcome a catalytic limitation, as has been generally assumed. Rather, their increased concentration may reflect an expanded mitochondrial population within the cell which compensates for the

Table 3. *Stereological comparison of mitochondrial populations in red and white myotomal muscle from goldfish acclimated to 5 °C and 25 °C. Values are expressed as mean $\pm$ S.E.M.*

Fibre type	Stereological parameter	Acclimation temperature		
		25 °C	5 °C	
Red muscle	Volume fraction (V_v in %)	6.3 ± 1.3	18.1 ± 1.3	$P < 0.005$
	Surface density (S_v in $\mu m^2 \mu m^{-3}$)	0.32 ± 0.08	0.80 ± 0.07	$P < 0.005$
	Specific surface (S/V in $\mu m^2 \mu m^{-3}$)	5.02 ± 0.41	4.47 ± 0.35	n.s.
Sample size				
No. of animals		4	4	
Area of sections sampled (μm^2)		593000	633000	
Approx. no. fibres in area		277	749	
White muscle	Volume fraction (V_v in %)	0.8 ± 0.1	1.5 ± 0.2	$P < 0.01$
	Surface density (S_v in $\mu m^2 \mu m^{-3}$)	0.05 ± 0.01	0.08 ± 0.01	n.s.
	Specific surface (S/V in $\mu m^2 \mu m^{-3}$)	6.28 ± 1.46	5.32 ± 0.63	n.s.
Sample size				
No. of animals		3	3	
Area of sections sampled (μm^2)		390000	880000	
Approx. no. fibres in area		66	246	

Volume fraction (volume density, V_v) is the volume fraction of mitochondria in the test volume; surface density (S_v) is the surface area of mitochondria in the test volume; specific surface (S/V) is the surface area of mitochondria per volume of mitochondria. Single obliquely transverse sections of each tissue block were used in the analysis. Sampling design for stereological analysis of muscle tissue requires sectioning at oblique angles. Consequently, the relationship between area sampled and number of fibres shown above should not be construed as an accurate reflection of mean fibre size. (Data are from S. Tyler & B. D. Sidell, unpublished.)

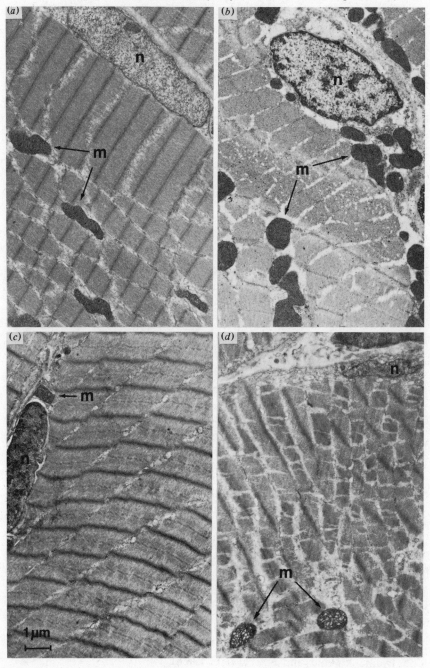

Fig. 5. Representative electron micrographs of comparable areas of red and white skeletal muscle from goldfish acclimated to 5 °C and 25 °C. (*a*: red, 25 °C; *b*: red, 5 °C; *c*: white, 25 °C; *d*: white, 5 °C). Note the increase in mitochondrial profiles in tissues from cold-acclimated animals. c, capillary; m, mitochondria; n, nucleus. (S. Tyler & B. D. Sidell, unpublished.)

impact of temperature on the exchange of adenylates and delivery of metabolites to the mitochondrial compartment. Johnston & Maitland (1980) have also correctly noted that this response additionally will enhance the rate of ATP provision from mitochondria to the contractile apparatus of the muscle cell.

Quantitative changes in the concentration of intracellular organelles (and their constituent enzymes) may be a fairly widespread response of cellular acclimatisation to chronic environmental change. It is encouraging to note that Campbell & Davies (1978) have reported increases in hepatic mitochondrial populations at cold acclimation temperature in the blenny and that Penney & Goldspink (1980) have recently described similar morphometric changes in goldfish muscle sarcoplasmic reticulum. Penney & Goldspink suggest that the changes in sarcoplasmic reticulum are instituted to overcome a diffusion limitation on the rates of Ca^{2+} release and resequestration at cold acclimation temperatures.

Conclusions

Individual cells within multicellular organisms have the ability to adapt to long-term environmental change by varying the concentration of specific enzymes and entire organelles. A critical feature which permits this mode of response is the dynamic flux of cellular constituents through processes of turnover. In fact, our current understanding suggests that the only component of the individual living cell that is not characterised by a continuous finite rate of turnover is the genetic material itself, DNA. The concentration of all other cellular macromolecules, at least potentially, may be altered by selective readjustment of their rates of synthesis and degradation.

By using examples from studies of metabolic compensation of fishes to temperature change I have attempted to illustrate the adaptive value of this 'quantitative strategy'. As in many other aspects of cellular biology, these examples reinforce the inviolate linkage between structure and function in biological systems. To appreciate metabolic adaptations of cells a thoughtful synthesis of both physical and chemical viewpoints is essential. By carefully combining information acquired through modern techniques in quantitative electron microscopy and biochemistry we can look forward to significant expansion in our understanding of cellular adaptations to environmental change.

The author's work is supported by grants from the (US) National Science Foundation (PCM 77-16838 and PCM 81-04112), American Heart Association, Maine Affiliate, and NATO (RG 136.8).

References

Caille, J. P. & Hinke, J. A. M. (1974). The volume available to diffusion in the muscle fiber. *Canadian Journal of Physiology and Pharmacology*, **52**, 814–28.

Campbell, C. M. & Davies, P. S. (1978). Temperature adaptation in the teleost, *Blennius pholis*: changes in enzyme activity and cell structure. *Comparative Biochemistry and Physiology*, **62B**, 165–7.

Dean, R. J. (1978). *Cellular Degradative Processes*. London: Chapman & Hall.

Fox, H. M. (1955). The effect of oxygen on the concentration of haem in invertebrates. *Proceedings of the Royal Society of London, Series B*, **143**, 203–14.

Goldberg, A. L. & Dice, J. F. (1974). Intracellular protein degradation in mammalian and bacterial cells. *Annual Review of Biochemistry*, **43**, 835–69.

Goodridge, A. G. (1975). Hormonal regulation of the activity of the fatty acid synthesizing system and of the malic enzyme concentration in liver cells. *Federation Proceedings*, **34**, 117–23.

Hazel, J. R. & Prosser, C. L. (1974). Molecular mechanisms of temperature compensation in poikilotherms. *Physiological Reviews*, **54**, 620–77.

Hochachka, P. W. & Somero, G. N. (1973). *Strategies of Biochemical Adaptation*. Philadelphia: W. B. Saunders.

Johnston, I. A. & Maitland, B. (1980). Temperature acclimation in crucian carp, *Carassius carassius* L.; morphometric analyses of muscle fiber ultrastructure. *Journal of Fish Biology*, **17**, 113–25.

Jones, P. L. & Sidell, B. D. (1982). Metabolic responses of striped bass (*Morone saxatilis*) to temperature acclimation. II. Alterations in metabolic carbon sources and distributions of fiber types in locomotory muscle. *Journal of Experimental Zoology*, **219**, 163–71.

Kushmerick, M. J. & Podolsky, R. J. (1969). Ionic mobility in muscle cells. *Science*, **166**, 1297–8.

Lewin, B. (1980). *Gene Expression*, vol. 2, *Eucaryotic Chromosomes*. New York: Wiley-Interscience.

Mahler, M. (1978). Diffusion and consumption of oxygen in the resting frog sartorius muscle. *Journal of General Physiology*, **71**, 533–57.

Meek, G. A. & Elder, H. Y. (1977). *Analytical and Quantitative Methods in Microscopy. SEB Seminar Series 3*. London: Cambridge University Press.

Newsholme, E. A. & Start, C. (1973). *Regulation in Metabolism*. New York: Wiley.

Penney, R. K. & Goldspink, G. (1980). Temperature adaptation of sarcoplasmic reticulum of goldfish. *Journal of Thermal Biology*, **5**, 63–8.

Schimke, R. T. & Doyle, D. (1970). Control of enzyme levels in animal tissues. *Annual Review of Biochemistry*, **39**, 929–76.

Shaklee, J. B., Christiansen, J. A., Sidell, B. D., Prosser, C. L. & Whitt, G. S. (1977). Molecular aspects of temperature acclimation in fish: contributions of changes in enzyme activities and isozyme patterns to metabolic reorganization in the green sunfish. *Journal of Experimental Zoology*, **201**, 1–20.

Sidell, B. D. (1977). Turnover of cytochrome *c* in skeletal muscle of green sunfish (*Lepomis cyanellus* R.) during thermal acclimation. *Journal of Experimental Zoology*, **199**, 233–50.

Sidell, B. D. (1980). Responses of goldfish (*Carassius auratus* L.) muscle to acclimation temperature: alterations in biochemistry and proportions of different fiber types. *Physiological Zoology*, **53**, 98–107.

Somero, G. N. (1978). Temperature adaptation of enzymes: biological optimization through structure–function compromises. *Annual Review of Ecology and Systematics*, **9**, 1–29.

Terjung, R. L., Winder, W. W., Baldwin, K. M. & Holloszy, J. O. (1975). Effect of exercise on the turnover of cytochrome *c*. *Journal of Biological Chemistry*, **248**, 7404–6.

Waterlow, J. C., Garlick, P. J. & Millward, D. J. (eds.) (1978). *Protein Turnover in Mammalian Tissues and in the Whole Body*. Amsterdam & London: North-Holland.

Wilkinson, C. F. (1979). The use of insect subcellular components for studying xenobiotics. In *Xenobiotic Metabolism*. In vitro *Methods*, ed. G. D. Paulson, D. S. Frear & E. P. Marks, pp. 249–84. Washington, DC: American Chemical Society.

Wilson, F. R. (1973). Enzyme changes in the goldfish (*Carassius auratus* L.) in response to temperature acclimation. I. An immunochemical approach. II. Isozymes. PhD Thesis, University of Illinois.

Yancy, P. H. & Somero, G. N. (1978). Temperature dependence of intracellular pH: its role in the conservation of pyruvate apparent K_m values of vertebrate lactate dehydrogenases. *Journal of Comparative Physiology*, **125**, 129–34.

Zeldin, M. H., Skea, W. & Mattson, D. (1973). Organelle formation in the presence of a protease inhibitor. *Biochemical and Biophysical Research Communications*, **52**, 544–9.

IAN A. JOHNSTON

Cellular responses to an altered body temperature: the role of alterations in the expression of protein isoforms

Temperature directly affects the stability of weak bonds that contribute to the tertiary structure of proteins and their interaction with ligands. Where animals have become adapted to a particular body temperature over long periods of evolutionary time the structural and functional characteristics of their proteins are often optimised for that temperature. For example, myosins isolated from cold-water fish species have high ATPase activities at low temperatures but are relatively unstable at higher temperatures and readily undergo aggregation and denaturation on storage (Connell, 1961; Johnston & Walesby, 1977).

Many animals, however, are able to function with a variable body temperature. For example, temperate freshwater fishes may experience a difference in temperature of 25–30 deg C between summer and winter. Similarly, some mammals drop their 'set' temperature down to a very low level ($\sim$ 2 °C) during hibernation. The adaptations which occur at the cellular level to accommodate the effects of temperature changes are very complex and as yet poorly understood. They include, or are manifest, as changes in metabolic rate, intracellular pH, ion and metabolite concentrations, membrane composition and alterations in gene expression (see Hazel & Prosser, 1974; Willis, 1979).

Many proteins can exist in a variety of different forms (isoforms). Each has a distinct primary structure, slightly different functional properties and is coded for by a separate gene. Although genes for all the different forms are present in the nucleus of each cell their expression varies during development, between cell types, and according to the physiological state of the organism. The present article considers the extent to which alterations in isoform composition are important in the cellular responses to an altered body temperature.

Terminology

It is important to distinguish between phenotypic variation arising from the expression of genes coding for different functional isoforms of individual proteins, and that due to genetic variation in which conservative mutations may occur in one or more of these different genes. Protein variants arising from genetic variation within populations or races of a species are called alloforms (see Powers, this volume, p. 227).

Examples of protein polymorphism and isoenzymes

Most of the myofibrillar proteins of skeletal muscle can occur in a number of different isoforms (Perry, 1979; Perry & Dhoot, 1981). For example, myosin, the principal constituent of the thick filament, is made up of heavy (2×200000 daltons) and light ($4 \times 16000-25000$ daltons) chain subunits. In vertebrates there is evidence for a large number of isoforms of both heavy and light chain subunits (Fig. 1). Different heavy or light chain isoforms can be recognised on the basis of changes in mobility during one- or two-dimensional electrophoresis, susceptibility to proteolytic digestion, amino acid sequence and immunological specificity (Gauthier & Lowey, 1979; d'Albis, Pantaloni & Bechet, 1979; Dhoot & Perry, 1979). There is evidence that the expression of subunits during development is subject to complex and co-ordinated genetic control. For example, in developing rabbit muscle there is evidence for the sequential expression of three foetal forms of myosin heavy chains (Whalen et al., 1981). However, while differences in ATPase activity have been demonstrated between native myosin isoenzymes present in fast, slow and cardiac muscles (Gergely et al., 1965), the functional significance of the large number of possible myosin isoforms is unclear.

A variety of experimental procedures can affect the expression of myosin genes in adult muscles including denervation, changes in functional demands and activity patterns (reviewed by Jolesz & Sreter, 1981). There is also evidence that some 'physiological' stimuli such as exercise (Saltin et al., 1977)

Fig. 1. Some of the polymorphic forms of myosin so far demonstrated in vertebrate striated muscle. Light-chain heterogeneity exists in different myosin molecules from foetal, skeletal and cardiac muscles (1–3). Similarly, different heavy-chain isoforms are also present in these muscle types. The degree to which light- and heavy-chain isoforms can associate into different hexamers within various muscle types is not established. For original sources of information see: 1, Jolesz & Sreter (1981); 2, Perry & Dhoot (1981) and references therein; 3, Hamoir et al. (1980); 4, Whalen et al. (1981); 5, Libera et al. (1980); 6, Hoh et al. (1979).

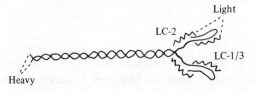

and altered hormone levels (Johnson et al., 1980) can affect the type of myosin synthesised.

Many aquatic poikilotherms show a partial or complete compensation of locomotory activity following acclimation from summer to winter temperatures (Bullock, 1955). Similarly, the diaphragm and cardiac tissues of hibernating mammals are able to operate at temperatures ($\sim 2\,°C$) which would result in loss of contractility in those muscles of other non-hibernating mammals (see Willis, 1979). Whether an altered expression of myosin subunits is involved in the conservation of contractile function at different body temperatures remains unexplored.

An example of a complex isoenzyme system which has been extensively studied in relation to temperature acclimation is L-lactate dehydrogenase (LDH) (EC 1.1.1.27). This is an enzyme catalysing the interconversion of lactate and pyruvate close to equilibrium in the presence of a co-enzyme nicotinamide adenine dinucleotide. In mammalian tissues there are two major structural genes for LDH coding for the M and H chains, together with a variety of other genes which can be expressed in some tissues at a certain stage of development. Each LDH molecule is composed of four subunits. Thus random association of the M and H polypeptides gives rise to five tetrameric isoenzymic forms: M_4, M_3H_1, M_2H_2, M_1H_3 and H_4 (Markert, 1963). In the early stages of embryonic development *M* and *H* genes are expressed at the same rate, but as differentiation proceeds the distribution of different isoenzymes begins to resemble that of the adult tissue. The functional significance of the different LDH isoenzymes has been the subject of much speculation. One suggestion is that the greater degree of pyruvate inhibition of the H_4 isoenzyme favours lactate oxidation and the channelling of pyruvate through the tricarboxylic acid cycle. Certainly there is a correlation between the aerobic character of a tissue and the presence of the H_4 or 'heart type' isoenzyme. For example, rat heart LDH contains 78% of H polypeptide chains compared with only 11% for leg muscles (Fine, Kaplan & Kuftinec, 1963). Electrophoretic patterns of fish LDHs are even more complex, suggesting the presence of still other genes in addition to those coding for the M and H polypeptides. For example, the LDH *E* locus is found in teleosts. The products of this gene, which is thought to have arisen from a duplication of the *H* gene around the time of the adaptive radiation of the teleosts, is expressed in nervous tissues, particularly photoreceptor cells (Holbrook et al., 1975). The *F* locus occurs in gadoids and is expressed in the liver (Fig. 2).

In addition, electrophoretically defined allelic variants of LDH are extremely widespread and in some cases gene frequency can be correlated with latitude (see Powers, this volume, p. 227). In the case of *Fundulus heteroclitus* it has been suggested that these directional changes in gene variation are maintained by the selective pressures of the steep thermal gradient found along the animal's distribution range (Powers & Place, 1978).

Where conclusions about an altered isoenzyme composition are based on electrophoretic evidence alone, the complexity of the patterns obtained can easily lead to erroneous conclusions. In order to understand the adaptive significance of any alterations in isoenzyme expression it is also necessary to consider structural and kinetic data. Unfortunately, in this context many of the published data are incomplete.

Isoforms or alloforms?

The studies by Wilson, Whitt & Prosser (1973) and Shaklee *et al.* (1977) demonstrate the importance of distinguishing between selective expression of different functional forms of a given protein (isoforms) and genetic variation (different alloforms). This can be a problem when sampling from small numbers of a wild population of fish or amphibians where there is often extensive heterozygosity at a particular gene locus. This point is well illustrated by Shaklee *et al.* (1977). They investigated the electrophoretic properties of phosphoglucomutase (PGM), a glycolytic enzyme, from the muscle of the green sunfish *Lepomis cyanellus*. Fig. 3 is taken from their paper and is an electrophoretogram of PGM from four fish that appears to show a more anodic component in 25 °C-acclimated relative to 5 °C-acclimated fish. However, when a larger number of fish are studied (right) it becomes apparent that the two bands correspond to the homozygous forms of a genetic polymorphism at a single PGM locus in the population studied. In some cases

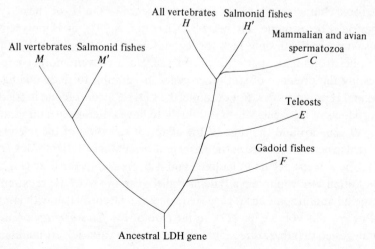

Fig. 2. Evolutionary relationships of vertebrate lactate dehydrogenase (LDH) genes.

the heterozygous form, consisting of the 'fast' and 'slow' migrating alleles, is observed regardless of the acclimation temperature of the fish (Fig. 3).

Body temperature and pH

Studies with a variety of poikilotherms have shown that both blood pH and intracellular pH vary inversely with body temperature (see Reeves, 1977). The significance of this pattern of acid–base regulation has been variously ascribed to a requirement to maintain constant either 'relative alkalinity' or the fractional charge state of histidine groups on proteins. The latter constitute the major ionizable group of proteins in the physiological pH range and thus dominate the buffering capacity of the cytoplasm (Reeves, 1977). Although absolute pH values of the blood and tissues vary considerably between species at a particular temperature, the rate of change of pH with temperature (dpH/dT) is remarkably constant and is close to -0.017 U per deg C. This value corresponds closely to the dpH/dT for histidine imidazole groups. Terrestrial poikilotherms appear to regulate their blood pH by changes in P_{CO_2} brought about by adjustments in the rate of ventilation. Aquatic poikilotherms, on the other hand, rely on a variety of ion-exchange

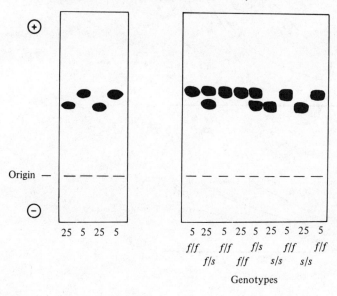

Fig. 3. Muscle phosphoglucomutase isoenzymes from green sunfish acclimated to either 5 °C or 25 °C. On the left, electrophoresis of samples from two 5 °C- and two 25 °C-acclimated fish apparently show qualitative differences in isoenzyme pattern. However, when samples from a larger number of fish are run (right) the existence of three genotypes becomes apparent. (From Shaklee et al., 1977.)

processes at the gills in order to adjust their body pH values to changes in environmental temperature.

The kinetic properties of enzymes measured *in vitro* at different temperatures are markedly affected if the effect of temperature on pH is not taken into account in the assays (Wilson, 1977; Yancey & Somero, 1978). For example, the effects of temperature on the apparent K_m values for pyruvate of muscle M_4 LDHs was markedly different between experiments in which pH was maintained constant and those in which pH was allowed to vary with temperature (Fig. 4). Only when buffers are chosen that have dpH/dT close to that found *in vivo* (i.e. -0.017 per deg C), are similar K_m values obtained for the different species. These data nicely illustrate the importance of using physiologically realistic assay conditions in comparative enzymology.

Other factors that are likely to be important include the range of substrate and co-factor concentrations, ionic strength, and whether the enzyme under study is normally bound to membranes or other proteins *in vivo*. In evaluating

Fig. 4. Effects of temperature on apparent Michaelis constants (K_m) for pyruvate of M_4 lactate dehydrogenase for various vertebrate species. (*a*) Assays carried out in phosphate buffer, pH 7.4 at all temperatures; (*b*) assays carried out in imidazole-HCl buffer with pH allowed to vary with temperature in a physiological manner. A, bluefin tuna; B, mudsucker; C, *Potamotrygon* sp.; D, rabbit. (Adapted from Yancey & Somero, 1978.)

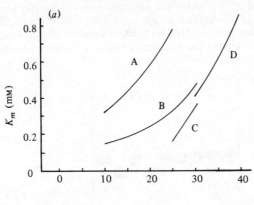

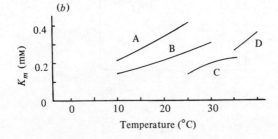

Temperature (°C)

the literature on temperature acclimation it is worth remembering that many of the early and some recent studies on enzyme function have paid little attention to these factors.

Cellular responses to seasonal temperature change

Homeotherms have evolved a number of strategies to cope with the problems of maintaining a constant body temperature during cold seasons. These include behavioural responses, better insulation in the form of subcutaneous fat deposits and extra fur, increased metabolic heat production, and hibernation. The relative importance of the different adaptations employed varies with latitude, the severity of the cold season, and body size. For example, large arctic mammals such as polar bears rely extensively on an effective insulation and do not show any metabolic response to cooling until the outside air temperature drops below -50 °C (Irving & Krog, 1955). On the other hand, small mammals from cool-temperate and subarctic zones rely extensively on non-shivering thermogenesis to maintain a constant body temperature, and consequently exhibit a number of compensatory adaptations at a cellular level. Perhaps the most dramatic response to cold winters is seen in hibernating mammals, where the regulated body temperature is reduced to a very low level, usually around 2 °C. During hibernation, processes such as the maintenance of ion gradients across membranes, the activity of temperature-sensitive neurones, and the contractility of the heart and diaphragm must all be maintained, albeit at a lower level, in spite of drastically lowered body temperature. Cellular adaptations underlying hibernation are very complex and as yet relatively unexplored (see Willis, 1979).

Poikilotherms show a variety of both cellular and behavioural adaptations to cold, including torpor and varying degrees of adjustment of metabolic rate and activity (Fig. 5).

Several different solutions to the problem of decreased enzyme activity with reduced body temperature seem to have arisen during the course of evolution. These can be broadly stated as follows:

1. In the case of allosteric enzymes, compensatory increases in metabolic rate may arise from changes in intracellular concentrations of H^+ and other ions, substrates, inhibitors and activators independently of any alterations in enzyme concentration or kinetics (Freed, 1971; Walesby & Johnston, 1980). Similarly, many enzymes are bound to membranes and lipid–protein interactions are important in determining their activity. The role of altered membrane composition in stabilising enzyme function is discussed by Cossins elsewhere in this volume.

2. Another solution to a variable body temperature is the selection of protein variants which represent a compromise between the optimal kinetic

forms for the high and low ends of the temperature range experienced. A good example is provided by the myofibrillar ATPase activity of *Fundulus heteroclitus*, a highly eurythermal species inhabiting the salt marshes of the Eastern Atlantic seaboard of the USA. Both the concentration dependence of ATPase on Ca^{2+} and catalytic activity of the ATPase show a very low temperature dependence, relative to other fish, across the range 12–35 °C (Johnston *et al.*, 1982).

3. A third mechanism for increasing enzyme activity at low temperatures is to increase the concentration of enzymes in the cell. This seems to be particularly important for adjusting the rate of aerobic metabolism (see Sidell, this volume).

4. There have been few studies of the effects of temperature on the functional properties of purified isoforms of a given protein. In animals with a variable body temperature it is possible that of those protein isoforms expressed in a given cell type some are better suited for function over a particular part of the animal's temperature range than others. If this is the case then it provides a way of maintaining the activity and regulatory

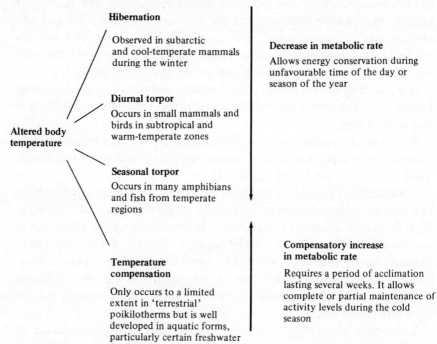

Fig. 5. Various metabolic strategies for coping with cold stress that involve a reduction in body temperature.

properties of proteins in response to rapid changes in temperature without the necessity for any change in gene expression.

5. It has also been suggested that differential expression of protein isoforms provides another important mechanism for regulating enzyme function following temperature acclimation or hibernation (Hochachka & Somero, 1973). One possibility is that there are temperature-specific isoenzymes in some animals which are best suited for function at a particular temperature (e.g. lower Q_{10}, or increased affinity for substrate: see Hochachka & Somero (1973) for discussion). Another possibility is that body temperature might influence the relative proportions of protein subunits of complex isoenzymes which are coded for at multiple gene loci. For example, a lowered body temperature might increase the relative proportion of the M-type LDH polypeptide in a tissue. Such changes in phenotypic expression may reflect a shift in the relative importance of a particular metabolic pathway by any of the other mechanisms outlined above. A final possibility is that different tissue-specific isoenzymes, e.g. M_4 versus H_4 LDHs, have somewhat different thermal properties.

Hibernation

The control of carbon flow through a particular metabolic pathway is largely determined by certain rate-limiting enzymes, the activity of which can be modulated by various allosteric effectors (activator or inhibitor molcules) (see Newsholme & Paul, this volume). Many regulatory enzymes occur in isoenzymic forms and might be expected to be a prime focus for evolutionary change in animals with variable body temperatures. The properties of one such enzyme, pyruvate kinase (PK), have been compared in hibernating and awake ground squirrels (Behrisch, 1974) and Little Brown Bats (*Myotis lucifugus*) (Borgman & Moon, 1976; Moon & Borgman, 1976). Both of these studies showed some interesting changes in the kinetics of the enzyme with hibernation. These included changes in apparent K_m-temperature relationships, shifts in the slopes and shapes of Arrhenius plots and altered sensitivity to various allosteric effectors (Fig. 6). For example, liver PK is much more sensitive to activation by fructose diphosphate in hibernating than in awake bats (Moon & Borgman, 1976).

There is some evidence that these changes may reflect an altered isoenzyme composition. Moon & Borgman (1976) found electrophoretic evidence for changes in isoenzyme composition in both muscle and livers of hibernating relative to awake bats. In the case of the liver enzyme, five electrophoretically distinct forms were observed in high-speed supernatants from both awake and hibernating bats although their relative positioning was slightly different. The number of electrophoretically distinct forms of PK in muscle was found to be much less for hibernating than active bats (Borgman & Moon, 1976). However, an unequivocal demonstration that altered isoenzyme expression may change the properties of key regulatory enzymes during hibernation must await studies with purified isoenzymes.

Temperature compensation in poikilotherms

The biology of temperature compensation is exceedingly complex and presents a number of problems of experimental design. For example, the metabolic responses to temperature acclimation will vary according to the length of acclimation, both the absolute and rates of change of temperature studied, the time of year the experiments are started, and with other factors

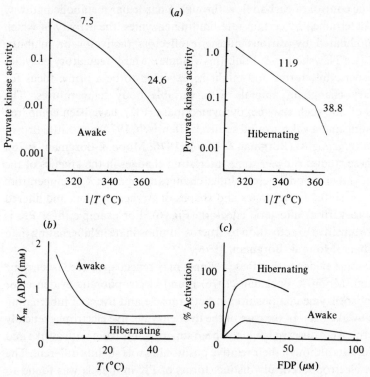

Fig. 6. The effect of temperature on various properties of pyruvate kinase from awake and hibernating Little Brown Bats (*Myotis lucifugus*). (*a*) Arrhenius plots of the activity of the liver enzyme at saturating ADP (4 mM) and subsaturating phosphoenol pyruvate (0.2 mM) concentrations. Calculated activation energies are in the units kcal mol^{-1}. (*b*) Estimated K_m (ADP) values as a function of temperature for the muscle enzyme. (*c*) The effects of fructose-1,6-diphosphate (FDP) on the activity of liver pyruvate kinase at 4 °C. (From Borgman & Moon (1976) and Moon & Borgman (1976), with permission.)

such as dissolved oxygen levels and photoperiod. The metabolic responses to temperature acclimation also vary between tissues and cell types. For example, Shaklee et al. (1977) acclimated groups of green sunfish to either 5 °C or 25 °C. While mitochondrial enzyme activities were generally observed to increase with cold acclimation the activities of various glycolytic enzymes increased in the brain and yet decreased somewhat in muscle (Fig. 7).

A general finding is that cold acclimation in many temperate fishes requires an acclimation period lasting several weeks and is associated with an increase in cellular respiration rate, a change in glycolytic flux and alterations in the importance of the pentose shunt pathway (Kanungo & Prosser, 1959; Hochachka & Hayes, 1962; Smit, Van Den Berg & Kijn-Den Hartog, 1974).

Quantitative alterations in gene expression appear to be a major factor in restructuring intermediary metabolism following temperature acclimation. For example, increased tissue oxygen consumption rates in cold-acclimated fish are associated with an increase in the number of mitochondria per cell (Johnston & Maitland, 1980; Johnston, 1982a), and an improvement in muscle capillary supply (Fig. 8). However, as with hibernation, the specific role of altered isoenzyme expression is much less clear.

Some investigators have reported changes in isoenzyme composition following temperature acclimation while others have not – even when the studies have been carried out on the same species under approximately

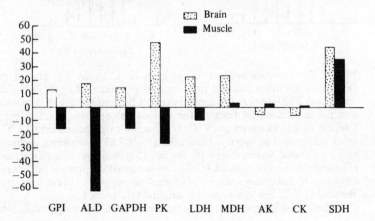

Fig. 7. Relative changes in the activity of some enzymes of energy metabolism in the brain and skeletal muscle of green sunfish acclimated to either 5 °C or 25 °C. The results are expressed as the percentage change in activity of 5 °C-acclimated relative to 25 °C-acclimated fish. GPI, glucose phosphate isomerase; ALD, alodolase; GAPDH, glycerophosphate dehydrogenase; PK, pyruvate kinase; LDH, lactate dehydrogenase; MDH, malate dehydrogenase; AK, adenylate kinase; CK, creatine phosphokinase; SDH, succinic dehydrogenase. (From Shaklee et al., 1977.)

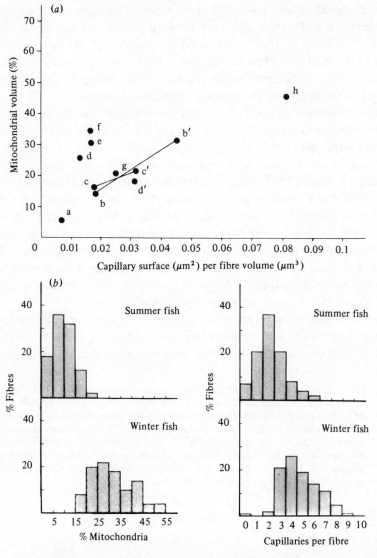

Fig. 8. Temperature acclimation in carp (2 °C vs 28 °C): altered gene expression for mitochondrial proteins and associated changes in capillary supply. (a) Relation between capillary supply per unit volume of slow muscle and the mitochondrial fraction for various fish species. a, Chimaera; b, Crucian carp acclimated to 28 °C; b′, Crucian carp acclimated to 2 °C; c, tench acclimated to hypoxia conditions (P_{O_2} 1.7 kPa); c′, tench acclimated to air-saturated water; d and d′, dogfish (*Scyliorhinus canicula*); e, shark (*Etmopterus spinax*); f, shark (*Galeus melastomus*); g, juvenile eel (*Anguilla anguilla*); h, European anchovy (*Engrasulosus mordax*). (From Johnston & Bernard (1982). See this paper for original references.)

(b) Left: Histograms showing the frequency distribution of mitochondrial comparable conditions (fish: Hochachka, 1965; Hochachka & Lewis, 1971; Moon & Hochachka, 1971; Wilson et al., 1973; Tsukuda, 1975; Somero, 1975; Shaklee et al., 1977; Amphibia: Enig, Ramsay & Eby, 1976; Tsugawa, 1980). For example, goldfish muscle and/or liver LDH isoenzyme composition has been variously reported as changing (Hochachka, 1965; Yamawaki & Tsukuda, 1979; Tsukuda & Yamawaki, 1980) or remaining unaltered (Wilson et al., 1973; Shaklee et al., 1977) following acclimation to cold temperatures (2–8 °C) or to warm (15–25 °C) temperatures. Much of this apparent contradiction hinges on what the different investigators have considered to constitute altered isoenzyme composition. It would appear that the 'on–off' switching of particular genes resulting in products which are only observed over a particular temperature range is exceedingly rare (see below). In contrast, there would appear to be several well-documented examples of changes in the relative proportions of different subunits in complex isoenzymes following acclimation. It should be pointed out that many of the above studies have failed to make any quantitative analysis of the relative proportions of different isoenzymes and have based their arguments entirely on qualitative assessment of badly overloaded electrophoretograms.

When run on starch or polyacrylamide electrophoresis gels the most anodic migrating isoenzyme is the heart type H_4 (LDH1) and the most cathodic the skeletal muscle type M_4 (LDH5). Cold acclimation has been shown to result in a relative increase in the more cathodic isoenzymes in some tissues from such diverse animals as *Xenopus* (Tsugawa, 1980), goldfish (Hochachka, 1965) and bats (Brush, 1968). For example, brain and liver LDH from the Painted Frog, *Discoglossus pictus pictus*, show a decrease in LDH1, 2 and 3 following cold-acclimation (8 °C versus 18 °C) and an increase in LDH4 and 5 (De Costa, Alonso-Bedate & Fraile, 1981). In some cases an increase in the cathodic muscle type LDH isoenzyme can be correlated with an increased ratio of pyruvate reduction to lactate oxidation at low temperatures (Tsukuda & Yamawaki, 1980). Tsugawa (1980) found that acclimation of *Xenopus laevis*, to either 14 or 25 °C did not change the temperature dependence of the apparent K_m of liver LDH for pyruvate. However, the H_4 and H_3M isoenzymes isolated by affinity chromatography had a higher Q_{10} for lactate oxidation than that of the M_4 isoenzyme. This, together with the relative decline in the amounts of H_4 and H_3M isoenzymes with cold acclimation, would also be expected to favour pyruvate reduction at low temperatures.

Fig. 8. (*cont.*)
volume in the slow muscle of carp acclimated to either winter (2 °C) or summer (28 °C) conditions. (From Johnston, 1982a.) Right: Histograms showing the frequency distribution of the number of capillaries per fibre in slow muscle of carp acclimated to either winter (2 °C) or summer (28 °C) conditions. (From Johnston, 1982a.)

Studies involving temperature acclimation in trout have also shown changes in the temperature dependence and kinetic properties of liver isocitrate dehydrogenase (Moon & Hochachka, 1972) together with an altered isoenzyme subunit composition. However, as with the results on PK from hibernating animals the available data are essentially correlative in nature and do not allow firm conclusions to be drawn about the thermal properties of specific gene products.

Little is known about the mechanism of regulation of changes in enzyme subunit composition with temperature acclimation, although there is some evidence that effects are mediated directly on cells. Tsugawa (1976) cultured two cell lines from *Xenopus* liver at a series of different temperatures and showed that the relative activity of anodic to cathodic isoenzymes decreased slightly in 15 °C relative to 25 °C cultures. The similarity of the changes of isoenzyme patterns *in vivo* and *in vitro* suggests an autonomous adaptation of the cells to local body temperature. However, other factors, such as tissue P_{O_2}, hormone levels and activity levels, could further modify the isoenzyme distribution pattern in the intact animal.

'On–off' synthesis of 'temperature-specific' isoenzymes?

There is little evidence supporting the presence of temperature-specific protein isoforms. One well-documented example is that of trout brain acetylcholinesterases (AChE) (Baldwin & Hochachka, 1970).

Rainbow trout (*Salmo gairdneri*) were acclimated to either 2, 12 or 17 °C for at least 4 weeks followed by isolation of a crude AChE preparation from the brain. AChE occurred in two distinct forms which migrated differently during electrophoresis in polyacrylamide gels. The anodic component was only found in brain extracts from fish acclimated at 17 °C whereas the cathodic band was only present in extracts from those acclimated at 2 °C. At the intermediate temperature (12 °C) both isoenzymic forms of AChE were present (Fig. 9b) (Baldwin & Hochachka, 1970). Although the isoenzymes expressed at 'high' and 'low' temperatures were found to have similar activation enthalpies, the temperature dependence of the apparent K_m values for acetylcholine was markedly different (Fig. 9a), the lowest K_m value coinciding with the acclimation temperature of each group of fish. Similar relationships between enzyme K_m and habitat temperature have been demonstrated for other fish species from diverse thermal environments (Yancey & Somero, 1978). Hochachka & Somero (1973) have suggested that such selective adjustments of enzyme K_m are of adaptive significance in providing some temperature independence of substrate and ligand binding properties. Similarly, Baldwin & Hochachka (1970) interpreted their results as evidence for the existence of 'warm' and 'cold' variants of AChE in trout brain, each kinetically optimised for the high and low ends of the fish's thermal range.

Temperature acclimation and the properties of myofibrillar ATPase

Thermal acclimation of myofibrillar ATPase activity is of particular interest since it can be directly related to behaviour and activity patterns. Furthermore, since myofibrils constitute more than 70% of the total protein in muscle cells, it is relatively straighforward to study the kinetics of the ATPase and isolate the individual components (myosin, actin, tropomyosin and the troponins). Measurements of the mechanical properties of skinned fibres (muscle cells from which the plasma membrane has been removed) also

Fig. 9. (a) The effect of temperature on the apparent K_m for acetylcholine of acetylcholinesterase isolated from trout brain. Fish were acclimated for several weeks to either 2 °C or 17 °C. (b) Electrophoretograms of acetylcholinesterase isoenzymes showing the different mobilities of the two forms found in 2 °C-, 12 °C- and 17 °C-acclimated fish. (From Baldwin & Hochachka, 1970.)

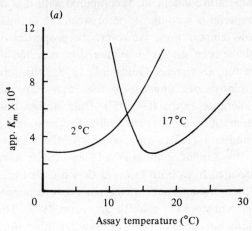

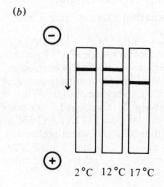

allows a correlation between the biochemistry of the various proteins and their physiological function (i.e. force production, rates of shortening). It is also clear that there are a number of functionally distinct forms of myosin (Fig. 1), tropomyosin and the troponin subunits I, T and C (Perry & Dhoot, 1981). In spite of the advantages of this system, temperature compensation of contraction has been little studied.

The rate of shortening of muscle is determined by the number of sarcomeres in series and by the effective rate of cross-bridge cycling, whereas tension is dependent on the number of myofilaments per unit cross-sectional area. Since both the length and spatial arrangement of filaments within a sarcomere are closely defined for a given muscle type, then it would appear that the mechanisms of adaptation might most appropriately operate by changing either cross-bridge characteristics or, for tension, the number of myofibres per cross-section of muscle (hypertrophy).

Some eurythermal species such as brook trout (Walesby & Johnston, 1981) and *Fundulus heteroclitus* (Johnston *et al.*, 1982) have evolved a myofibrillar ATPase which is relatively temperature-independent compared with that of other fishes and presumably represents a compromise between the optimal kinetic forms for high and low temperatures. However, the properties of goldfish myofibrillar ATPase have been shown to be altered by a period of temperature acclimation (Johnston, Davison & Goldspink, 1975; Johnston, 1979, 1980; Sidell, 1980). Goldfish are unusual in that they have an exceptionally wide thermal tolerance range (0–35 °C). Cold acclimation results in a relative increase in myofibrillar ATPase at low temperatures, a decrease in activation energy and an increased susceptibility of the ATPase to thermal denaturation (Fig. 10). Similar changes in ATPase activity have been reported for myofibrils isolated from both fast and slow muscle fibres (Johnston & Lucking, 1978), although somewhat varied acclimatory responses have been obtained between different stocks of goldfish (Johnston, 1979). The goldfish has been subject to many hundreds of years of intensive inbreeding in order to produce different strains. This genetic variation makes it a less than ideal species for temperature acclimation experiments. Unfortunately,

Fig. 10. (*a*) The effect of temperature on the $(Mg^{2+} + Ca^{2+})$-ATPase activity of natural actomyosin isolated from the fast muscle of 4 °C-acclimated (open circles) and 24 °C-acclimated (filled circles) goldfish (*Carassius auratus* L.). The activation enthalpies ($\Delta H\ddagger$) calculated from corresponding Arrhenius plots are also shown (units kcal mol^{-1}). Assay conditions: 40 mM imidazole, pH 7.4 at 10 °C, 30 mM KCl, 5 mM MgCl$_2$, 3.4 mM ATP, 0.1 mM CaCl$_2$.

(*b*) The effect of temperature on $(Mg^{2+} + EGTA)$-ATPase activity of actomyosin preparations isolated from cold- and warm-acclimated goldfish. Experimental conditions as shown above except that 5 mM EGTA replaced 0.1 mM CaCl$_2$. Values represent mean±S.E. of six preparations. (I. A. Johnston, unpublished results.)

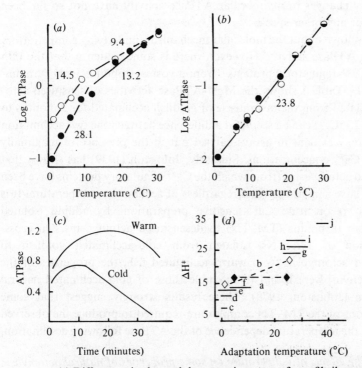

(c) Differences in thermal denaturation rates of myofibrils prepared from the fast muscle of goldfish acclimated to either 1 °C (open circles) or 26 °C (filled circles) for 2 months. Values represent mean ± S.E. of seven preparations. The plot shows ATPase activity remaining (μmol P_i per mg myofibrils per min) following pre-incubation of myofibrils (mg ml^{-1}) for various times at 37 °C in a medium of 0.05 M KCl, 40 mM Tris-HCl pH 7.5. (Data taken from Johnston et al., 1975).

(d) Relationship between the activation enthalpy ($\Delta H\dagger$) of fast muscle $Mg^{2+}Ca^{2+}$ myofibrillar ATPase activity and environmental temperature for various eurythermal and stenothermal fish. Vertical bars represent mean ± S.E. of at least eight preparations; horizontal bars correspond to the temperature range experienced by each species and the position of the point indicates the acclimation or adaptation temperature of the individuals sampled. (a) brook trout (*Salvelinus fontinalis*) (eurythermal species with an ATPase activity which appears to be a compromise between the optimal kinetic forms for the high and low ends of its thermal range. (b) goldfish acclimated to either 1 °C or 24 °C showing the change in thermodynamic activation parameters. (c–j) Somewhat more stenothermal species – (c) Icefish (*Champsocephalus gunnari*) (Antarctica), (d) *Notothenia neglecta* (Antarctica), (e) *Notothenia rossii* (Antarctic), (f) Short-horned sculpin (*Myoxocephalus scorpius*) (North Sea), (g) Abudefduf oxydon (Indo-Pacific Ocean), (h) *Pomatocentrus uniocellatus* (Indo-Pacific Ocean), (i) *Dascyllus carneus* (Indo-Pacific Ocean), (j) *Tilapia grahami* (Equatorial Hotsprings; Lake Magadi, Kenya) (from Johnston, 1982b).

acclimatory changes in myofibrillar ATPase activity have not so far been reported for any other species.

Little is known about the molecular mechanisms underlying compensatory changes in ATPase activity. However, there is some evidence that the thin filament Ca^{2+}-regulatory proteins (tropomyosins–troponins) play an important role. Table 1 shows the Mg^{2+}-ATPase activities of sensitised actomyosin isolated from the fast muscle of goldfish acclimated for 2 months to either 5 or 25 °C. It can be seen that a difference between the two acclimation groups is only evident in assays carried out in the presence of maximally activating Ca^{2+} concentrations. Similarly, Johnston (1979) has shown that desensitised actomyosins (from which the Ca^{2+}-regulatory proteins have been extracted) have similar activities regardless of acclimation temperature. It is possible to reconstitute Ca^{2+}-sensitive preparations by adding isolated tropomyosin–troponins (TM–TN) to desensitised actomyosins. On cross-hybridisation of TM–TNs isolated from cold-acclimated goldfish to desensitised actomyosin from warm-acclimated fish, the properties of the ATPase activity were found to resemble those of cold-acclimated natural actomyosin (Johnston, 1979). These results strongly suggest that some modifications of the TM–TN complex are required to produce the observed changes in the temperature dependence of the ATPase following acclimation.

Table 1. *The effects of temperature on some properties of natural actomyosin isolated from goldfish acclimated for 2 months to either 5 °C or 25 °C*

Assay temperature (°C)	ATPase activity (μmol P_i per mg myofibrils per minute)	
	Saturating Ca^{2+} concentration (10^{-4} M)	Trace Ca^{2+} concentration ($< 10^{-7}$ M)
5 °C-acclimated fish (8 hours daylight, 16 hours dark)		
0.5	0.22 ± 0.02	0.01 ± 0.00
26	2.40 ± 0.23	0.37 ± 0.05
25 °C-acclimated fish (16 hours daylight, 8 hours dark)		
0.5	0.07 ± 0.01	0.01 ± 0.00
26	2.44 ± 0.17	0.39 ± 0.04

Data represent mean $\pm$ s.e. for actomyosin preparations isolated from six fish at each acclimation temperature.

Currently, we are involved in a characterisation of the TM–TN complex. Initial results indicate that there are two forms of troponin subunit I in goldfish fast muscle differing slightly in molecular weight and electrophoretic mobility (unpublished results). One form predominates in the muscle of cold-acclimated fish and the other in warm-acclimated fish. It is not known whether a particular troponin I co-varies with isoforms of any of the other components of the TM–TN complex. Also, these results do not exclude acclimatory changes in myosin. Further studies on the immunological cross-reactivity and peptide mapping of myosin heavy and light chains would be required to resolve this point. However, the possibility that acclimation changes in ATPase can occur through modifications in the TM–TN complex alone is appealing since these proteins constitute a relatively small proportion of the myofibrillar mass and there is some evidence that they turn over faster than myosin.

General conclusions

There is a variety of evidence for altered isoenzyme composition following hibernation or temperature acclimation in poikilotherms. Many of these data are qualitative and correlative in nature. In order to understand their quantitative importance to metabolism and activity patterns it is essential that electrophoretic evidence of isoenzyme changes be accompanied by both structural and kinetic studies of the purified isoenzymes.

References

Baldwin, J. & Hochachka, P. W. (1970). Functional significance of isoenzymes in thermal acclimation. Acetylcholinesterase from trout brain. *Biochemical Journal*, **116**, 883.

Behrisch, H. W. (1974). Temperature and the regulation of enzyme activity in the hibernator. Isoenzymes of liver pyruvate kinase from the hibernating and non-hibernating Arctic ground squirrel. *Canadian Journal of Biochemistry*, **52**, 894–902.

Borgman, A. I. & Moon, T. W. (1976). Enzymes of the normothermic and hibernating bat, *Myotis lucifugus*: temperature as a modulator of pyruvate kinase. *Journal of Comparative Physiology*, **197**, 185–99.

Brush, A. H. (1968). Response of isozymes to torpor in the bat, *Eptesicus fuscus*. *Comparative Biochemistry and Physiology*, **27**, 113–20.

Bullock, T. H. (1955). Compensation for temperature in the metabolism and activity of poikilotherms. *Biological Reviews*, **30**, 31–42.

Connell, J. J. (1961). The relative stabilities of the skeletal muscle myosins of some animals. *Biochemical Journal*, **80**, 503–9.

d'Albis, A., Pantaloni, C. & Bechet, J. J. (1979). An electrophoretic study of native myosin isozymes and of their subunit content. *European Journal of Biochemistry*, **99**, 261–72.

De Costa, J., Alonso-Bedate, M. & Fraile, A. (1981). Temperature acclimation in amphibians: changes in lactate dehydrogenase activities and isoenzyme patterns in several tissues from adult *Discoglossus pictus pictus* (Otth.). *Comparative Biochemistry and Physiology*, **70B**, 331–9.

Dhoot, G. K. & Perry, S. V. (1979). Distribution of polymorphic forms of troponin components and tropomyosin in skeletal muscle. *Nature, London*, **278**, 714–18.

Enig, M., Ramsay, J. & Eby, D. (1976). Effect of temperature on pyruvate metabolism in the frog: the role of lactate dehydrogenase isoenzymes. *Comparative Biochemistry and Physiology*, **53B**, 145–8.

Fine, I. H., Kaplan, N. D. & Kuftinec, D. (1963). Developmental changes of mammalian dehydrogenases. *Biochemistry*, **2**, 116–21.

Freed, J. M. (1971). Properties of muscle phosphofructokinase of cold- and warm-acclimated *Carassius auratus*. *Comparative Biochemistry and Physiology*, **39B**, 747–64.

Gauthier, G. F. & Lowey, S. (1979). Distribution of myosin isoenzymes among skeletal muscle fibre types. *Journal of Cell Biology*, **81**, 10–25.

Gergely, J., Pragay, D., Scholz, A. F., Seidel, J. C., Sreter, F. A. & Thompson, M. M. (1965). Comparative studies on white and red muscle. In *Molecular Biology of Muscle Contraction*, ed. H. Kumagai & S. Ebashi, pp. 145–59. Amsterdam: Elsevier.

Hamoir, G., Geradin-Otthier, N. & Focant, B. (1980). Protein differentiation of the superfast swim bladder muscle of the toadfish *Opsanus tau*. *Journal of Molecular Biology*, **143**, 155–60.

Hazel, J. R. & Prosser, C. L. (1974). Molecular mechanisms of temperature compensation in poikilotherms. *Physiological Reviews*, **54**, 620–77.

Hochachka, P. W. (1965). Isoenzymes in metabolic adaptation of a poikilotherm: subunit relationships in lactic dehydrogenases of goldfish. *Archives of Biochemistry and Biophysics*, **111**, 96–103.

Hochachka, P. W. & Hayes, F. R. (1962). The effect of temperature acclimation on pathways of glucose metabolism in the trout. *Canadian Journal of Zoology*, **40**, 261–70.

Hochachka, P. W. & Lewis, J. K. (1971). Interacting effects of pH and temperature on the K_m values of fish tissue lactate dehydrogenases. *Comparative Biochemistry and Physiology*, **39B**, 925–33.

Hochachka, P. W. & Somero, G. N. (1973). *Strategies of Biochemical Adaptation*. Philadelphia: W. B. Saunders.

Hoh, J. F. Y., Yeoh, G. P. S., Thomas, A. W. & Higginbottom, C. (1979). Structural differences in the heavy chains of rat ventricular myosin isoenzymes. *FEBS Letters*, **97**, 330–4.

Holbrook, J. J., Liljas, A., Steindel, S. J. & Rossmann, G. (1975). Lactate dehydrogenase. In *The Enzymes*, vol. 11, *Oxidation–reduction*, part A, *Dehydrogenases (I), Electron Transfer (I)*, ed. P. D. Boyer, pp. 191–292. New York & London: Academic Press.

Irving, L. & Krog, J. (1955). Temperature of skin in the Arctic as a regulator of heat. *Journal of Applied Physiology*, **7**, 355–64.

Johnson, M. A., Mastaglia, F. L. & Montgomery, A. G. (1980). Changes in myosin light chains in the rat soleus after throidectomy. *FEBS Letters*, **110**, 230–5.

Johnston, I. A. (1979). Calcium regulatory proteins and temperature acclimation of actomyosin from a eurythermal teleost (*Carassius auratus* L.). *Journal of Comparative Physiology*, **129**, 163–7.

Johnston, I. A. (1981). Specializations of fish muscle. In *The Development and Specialisation of Skeletal Muscle*, SEB Seminar Series 7, ed. D. F. Goldspink, pp. 123–148. Cambridge University Press.

Johnston, I. A. (1982a). Capillarisation, oxygen diffusion distances and mitochondrial content of carp muscles following acclimation to summer and winter temperatures. *Cell and Tissue Research*, in press.

Johnston, I. A. (1982b). Physiology of muscle in hatchery raised fish. *Comparative Biochemistry and Physiology*, **73B**, 105–24.

Johnston, I. A. & Bernard, L. M. (1982). Routine oxygen consumption and characteristics of the myotomal muscle in tench: effects of long-term acclimation to hypoxia. *Cell and Tissue Research*, **227**, 161–77.

Johnston, I. A., Davison, W. & Goldspink, G. (1975). Adaptations in Mg^{2+}-activated myofibrillar ATPase activity induced by temperature acclimation. *FEBS Letters*, **50**, 293–5.

Johnston, I. A. & Lucking, M. (1978). Temperature induced variation in the distribution of different types of muscle fibre in the goldfish (*Carassius auratus*). *Journal of Comparative Physiology*, **124**, 111–16.

Johnston, I. A. & Maitland, B. (1980). Temperature acclimation in crucian carp: a morphometric analysis of muscle fibre ultrastructure. *Journal of Fish Biology*, **17**, 113–25.

Johnston, I. A., Sidell, B. D., Moerland, T. S. & Goldspink, G. (1982). Molecular plasticity in the myofibrillar protein complex of the Mummichog (*Fundulus heteroclitus*). *Journal of Muscle Research and Cell Motility*, in press.

Johnston, I. A. & Walesby, N. J. (1977). Molecular mechanisms of temperature adaptation in fish myofibrillar adenosine triphosphatases. *Journal of Comparative Physiology*, **119**, 195–206.

Jolesz, E. & Sreter, F. A. (1981). Development, innervation and activity-pattern induced changes in skeletal muscle. *Annual Review of Physiology*, **43**, 531–52.

Kanungo, M. S. & Prosser, C. L. (1959). Physiological and biochemical adaptation of goldfish to cold and warm temperature. II. Oxygen consumption of liver homogenate; oxygen consumption and oxidative phosphorylation of liver mitochondria. *Journal of Cellular and Comparative Physiology*, **54**, 265–74.

Libera, L. D., Sartore, S., Pieronbon-Bormioli, S. & Schiaffino, S. (1980). Fast-white and fast-red isomyosins in guinea pig muscles. *Biochemical and Biophysical Research Communications*, **96**, 1662–70.

Markert, C. L. (1963). Lactate dehydrogenase isozymes: dissociation and recombination of subunits. *Science*, **146**, 1329–30.

Moon, T. W. & Borgman, A. I. (1976). Enzymes of the normothermic and hibernating bat, *Myotis lucifugus*: metabolites as modulators of pyruvate kinase. *Journal of Comparative Physiology*, **107**, 201–10.

Moon, T. W. & Hochachka, P. W. (1972). Temperature and the kinetic analysis of trout isocitrate dehydrogenases. *Comparative Biochemistry and Physiology*, **42B**, 725–30.

Perry, S. V. (1979). The regulation of contractile activity in muscle. *Biochemical Society Transactions*, **7**, 593–617.

Perry, S. V. & Dhoot, G. K. (1981). Biochemical aspects of muscle development and differentiation. In *The Development and Specialisation of Muscle*, SEB Seminar Series 7, ed. D. F. Goldspink, pp. 51–64. Cambridge University Press.

Powers, D. A. & Place, A. R. (1978). Biochemical genetics of *Fundulus heteroditus* L. I. Temporal and spatial variation in gene frequencies of Ldh-B, Mdh-A, Gpi-B, and Pgm-A. *Biochemical Genetics*, **16**, 593–607.

Reeves, R. B. (1977). The interaction of body temperature and acid–base balance in ectotherms. *Annual Review of Physiology*, **39**, 559–86.

Saltin, B., Henriksson, J., Nygaard, E., Andersen, P. & Jansson, E. (1977). Fibre types and metabolic potentials of skeletal muscles in sedentary man and endurance runners. *Annals of the New York Academy of Sciences*, **301**, 3–29.

Shaklee, J. B., Christiansen, J. A., Sidell, B. P., Prosser, C. C. & Whitt, G. S. (1977). Molecular aspects of temperature acclimation in fish: contributions of changes in enzymic activities and isoenzyme patterns to metabolic reorganisation in the green sunfish. *Journal of Experimental Zoology*, **201**, 1–20.

Sidell, B. D. (1980). Responses of goldfish (*Carassius auratus* L.) muscle to acclimation temperature: alterations in biochemistry and proportions of different fibre types. *Physiological Zoology*, **53**, 98–107.

Smit, H., Van Den Berg, R. j. & Kijn-Den Hartog, I. (1974). Some experiments on thermal acclimation in the goldfish (*Carassius auratus* L.). *Netherlands Journal of Zoology*, **24**, 32–49.

Somero, G. N. (1975). The role of isozymes in adaptation to varying temperatures. In *Isozymes*, vol. 2, *Physiological Function*, ed. C. L. Markert, pp. 221–34. New York & London: Academic Press.

Tsugawa, K. (1976). Direct adaptation of cells to temperature: similar changes to LDH isozyme patterns by *in vitro* and *in situ* adaptations in *Xenopus laevis*. *Comparative Biochemistry and Physiology*, **55B**, 259–63.

Tsugawa, K. (1980). Thermal dependence in kinetic properties of lactate dehydrogenase from the African clawed toad, *Xenopus laevis*. *Comparative Biochemistry and Physiology*, **66B**, 459–66.

Tsukuda, H. (1975). Temperature dependency of the relative activities of the liver lactate dehydrogenase isozymes in goldfish acclimated to different temperatures. *Comparative Biochemistry and Physiology*, **52B**, 343–5.

Tsukuda, H. & Yamawaki, H. (1980). Lactate dehydrogenase of goldfish red and white muscle in relation to thermal acclimation. *Comparative Biochemistry and Physiology*, **67B**, 289–95.

Walesby, N. J. & Johnston, I. A. (1980). Temperature acclimation in brook trout muscle: adenine nucleotide concentrations, phosphorylation state and adenylate energy charge. *Journal of Comparative Physiology*, **139**, 127–33.

Walesby, N. J. & Johnston, I. A. (1981). Temperature acclimation of $Mg^{2+} \cdot Ca^{2+}$ myofibrillar ATPase from a cold-selective teleost, *Salvelinus fontinalis*: a compromise solution. *Experientia*, **37**, 716–18.

Whalen, R. G., Sell, S. M., Butler-Browne, G. S., Schartz, K., Bouveret, P. & Pinset-Harstrom, I. (1981). Three myosin heavy chain isoenzymes appear sequentially in rat muscle development. *Nature, London*, **292**, 805–9.

Willis, J. S. (1979). Hibernation: cellular aspects. *Annual Review of Physiology*, **41**, 275–86.

Wilson, F. R., Champion, M. J., Whitt, G. S. & Prosser, C. L. (1975). Isoenzyme patterns in tissues of temperature-acclimated fish. In *Isozymes*, vol. 2, *Physiological Function*, ed. C. L. Markert. New York & London: Academic Press.

Wilson, F. R., Whitt, G. S. & Prosser, C. L. (1973). Lactate dehydrogenase and malate dehydrogenase isoenzyme patterns in tissues of temperature-acclimated goldfish (*Carassius auratus* L.). *Comparative Biochemistry and Physiology*, **46B**, 105–16.

Wilson, T. L. (1977). Interrelations between pH and temperature for the catalytic rate of the M_4 isoenzyme of lactate dehydrogenase (EC 1.1.1.27) from goldfish (*Carassius auratus* L.). *Archives of Biochemistry and Biophysics*, **179**, 378–90.

Yamawaki, H. & Tsukuda, H. (1979). Significance of the variation in isoenzymes of liver lactate dehydrogenase with thermal acclimation in goldfish. I. Thermostability and temperature dependency. *Comparative Biochemistry and Physiology*, **62B**, 89–93.

Yancey, P. H. & Somero, G. N. (1978). Temperature dependence of intracellular pH: its role in the conservation of pyruvate apparent K_m values of vertebrate lactate dehydrogenases. *Journal of Comparative Physiology*, **125**, 129–34.

JOHN M. LITTLETON

Membrane reorganisation and adaptation during chronic drug exposure

Organisms which are exposed to fluctuations in environmental temperature, and which are unable to maintain their internal temperature constant in the face of these fluctuations, have effective cellular mechanisms of adaptation. One of the most basic mechanisms is the alteration in cell membrane lipid composition, or 'homeoviscous adaptation', which accompanies temperature fluctuations (see Cossins, this volume). The principle is, that in the face of a reduction in temperature, which would otherwise make the lipids of the membrane less fluid, the cell adapts by modifying the composition of its membrane towards those lipids which are more fluid, resulting in a compensation of the effect of temperature. The converse change, incorporation of lipids which are less fluid, occurs in response to an increase in environmental temperature. Most evidence suggests that the major change is in the degree of saturation of acyl chains of membrane phospholipids, but there is also evidence that changes in the proportion of cholesterol can be involved. Theoretically alterations in the proportions of phospholipids with different head groups could also be involved but there seems little direct evidence for this. This article examines the hypothesis, first formulated by Hill and Bangham in 1975, that cellular adaptation to certain drugs may occur by a mechanism similar to that of homeoviscous adaptation (Fig. 1).

The hypothesis rests on several assumptions which need to be briefly considered before the main evidence is reviewed. First is the assumption that some drugs act on cell membranes in a way which is similar to an alteration in temperature. Second is the assumption that the trigger for homeoviscous adaptation is the alteration in membrane characteristics rather than the temperature change *per se*. Thirdly, if such a mechanism is to be relevant to man, it must be assumed that the cells of homeothermic organisms, which do not need the capacity for homeoviscous adaptation to temperature, nevertheless retain this capacity and are able to employ it in response to drug-induced alterations in membrane viscosity. Certain other assumptions are required, but these three are of most importance and should now be discussed.

It is still in doubt as to whether any drugs produce their pharmacological effects in a manner which can be equated with an altered temperature of the cell membrane. There is much evidence that compounds such as general anaesthetics, alcohols and phenothiazine tranquillisers can enter cell membranes both expanding them and disrupting the normal packing of their lipids, i.e. making them more fluid (Seeman, 1972). However, even the most basic tenet of this hypothesis, that of drug entry into the membrane, has, for amphipathic molecules at least, recently been challenged (Singer, 1981) and it is by no means certain that any significant *functional* change in the membrane results from this fluidising effect at *pharmacological* concentrations of these agents. Despite these objections, the hypothesis remains that general depressants of the central nervous system inhibit cell function by entering the membrane lipid bilayer and subsequently disrupting some lipid–lipid or

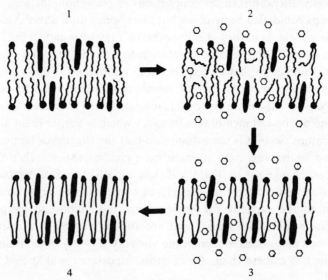

Fig. 1. Stages of membrane adaptation according to the homeoviscous drug-adaptation hypothesis. In (1) the normal membrane lipid organisation is shown with acyl chains projecting into the bilayer from phospholipid head groups and with cholesterol molecules interspersed between the phospholipids. In (2) a lipid-soluble drug is shown entering the membrane, disrupting the normal packing of the lipids and thus making the structure more fluid. In (3) the cell responds by incorporating lipids which are less fluid into its membrane. This is shown by the straighter acyl chains of the phospholipids, indicating a more saturated structure, and by the inclusion of more molecules of cholesterol into the membrane. This adaptation is to restore a normal level of fluidity to the membrane when the drug is present. The hypothesis predicts that, when the adapted membrane is tested in the drug-free state (4), it should be less fluid than the non-adapted membrane (1).

lipid–protein interactions which are important for normal cell function. This change can be detected by the incorporation of hydrophobic probes into the lipid bilayer and by the analysis of their spectroscopic properties (e.g. fluorescence polarisation, electron spin resonance). On addition of an alcohol or general anaesthetic, the motional properties of the probe are altered, almost always in a way similar to that shown by an increase in temperature, i.e. a 'fluidisation' (e.g. see Lenaz et al., 1979; Chin & Goldstein, 1981). Although this first assumption cannot be rigorously proved there is sufficient evidence to consider its implications, which include the homeoviscous drug-adaptation hypothesis.

The second assumption has been dealt with by Thompson (this volume). There is good evidence that the stimulus for homeoviscous adaptation is not simply the alteration in temperature, but that this change is 'sensed' by the cell as a result of alteration in the physical characteristics of some critical membrane component. Accordingly, drugs which also cause a change in the physical properties of this membrane component should also trigger homeoviscous adaptation.

The third assumption requires that mammalian cells should also be capable of homeoviscous adaptation. Homeoviscous adaptation to temperature in mammalian cells has been shown in culture (Henle & Dethlefsen, 1978), in hibernating mammals, where body temperature fluctuates markedly (Goldman, 1975) and in mammalian cells of the extremities, where the local temperature fluctuates in response to that of the environment (Cherqui et al., 1979). There seems no reason to suppose that mammalian cells have lost the capacity for homeoviscous adaptation even if they rarely require this capacity for physiological reasons. This argues that homeoviscous cellular adapation to drugs could be an important mechanism for drug tolerance in mammals and, specifically, in man.

Since most of the drugs which are thought to affect membrane lipids in this non-specific way are general depressants of the central nervous system, and since the one drug of this class which is administered chronically (to himself) by man is ethanol, much work has been directed towards the question of whether ethanol tolerance can be due to a homeoviscous adaptive change in the membrane lipids of the neurones of the human central nervous system. To this end a variety of experimental preparations have been used. I intend to review this evidence, beginning with simpler non-mammalian systems and ending with those which are more complex, such as experiments on intact mammals. To date very little work exists directly relevant to the question as to whether such changes occur in *human* brain. Where possible I shall mention evidence which relates to drugs other than ethanol but such evidence is relatively scanty.

Homeoviscous drug-adaptation in lower organisms

The first published experiments directly relevant to the homeoviscous drug-adaptation hypothesis seem to be those of Ingram (1976). In these experiments cultures of *Escherichia coli* were exposed to varying concentrations of alcohols of different chain lengths. Concentrations of alcohols were found which were initially inhibitory to bacterial growth but, after about 2 hours, bacterial growth resumed at the normal rate. Associated with this resumption was an alteration in the lipid composition of the bacterial membrane. It was argued that this alteration constituted a homeoviscous drug-adaptation of the membrane which conferred a degree of drug tolerance on the organism. Ingram has since extended this work in a series of papers (Ingram, Ley & Hoffman, 1978; Ingram, Buttke & Dickens, 1980) which suggests very strongly that *E. coli* adapts to alcohols by changing its lipid composition and that, in the absence of such alterations, drug tolerance does not occur.

This evidence, therefore, supports the homeoviscous drug-adaptation hypothesis. Indeed it goes a very long way towards proving that such a mechanism is important in bacteria. However, there is one very disturbing aspect to the work relating to the effects of alcohols of differing chain length. Ingram found that long-chain alcohols produced the predicted increase in the proportion of saturated lipids (to overcome the presumed fluidisation induced by the alcohols) but that short-chain alcohols, including ethanol, caused the incorporation of lipids which were less saturated. These results can be interpreted either as suggesting that short-chain alcohols do not cause membrane fluidisation, that short-chain alcohols do not cause homeoviscous adaptation (but produce bacterial tolerance by another mechanism), or that the membrane lipid alteration observed in response to all alcohols is not an example of homeoviscous adaptation and is probably not the direct cause of drug tolerance.

I confess that I am unable to decide which of these possibilities is most likely. Ingram (1976) favours the first, that short-chain alcohols do not fluidise bacterial membranes but produce the opposite result, although this is at variance with all the published evidence using physical techniques. In similar experiments on *Tetrahymena pyriformis*, Nandini-Kishore *et al.* (1979) showed that the presence of ethanol *did* fluidise cell membranes and that this organism also responded by incorporation of more fluidising lipids. Again this result is not consistent with the homeoviscous drug-adaptation hypothesis and it contrasts with earlier experiments from the same laboratory where it was shown that the presence of the general anaesthetic, methoxyflurane, in the environment did elicit an 'appropriate' incorporation of more viscous lipids into *T. pyriformis* membranes (Nandini-Kishore, Kitajima & Thompson, 1977).

The relevance of these results to adaptation in the mammalian central nervous system is debatable. In both the bacterial and *Tetrahymena* experiments the concentration of alcohol used was greater than the maximum tolerated by mammals (i.e. it would be lethal), and thus may have little relevance to the phenomena of tolerance and dependence in higher animals. It may be, for example, that, although incorporation into the membrane of lipids tending to decrease fluidity may have the opposite effect to that of ethanol, once these lipids have been incorporated ethanol has a greater disrupting effect on the membrane (Fig. 2). Alternatively, very high levels of ethanol may directly affect lipid metabolism. Such a hypothesis predicts that there should be a 'cross-over' point at which an appropriate response to increasingly high concentrations of ethanol would be incorporation of more fluid lipids into the cell membrane. Results consistent with this hypothesis have been found using human blood platelets as an experimental model (see p. 150) and such an effect would explain the discrepancies between the results obtained in both bacteria and *Tetrahymena* and the homeoviscous drug-adaptation hypothesis.

Such work as has been performed on other drugs does not really aid interpretation. The effect of methoxyflurane on *Tetrahymena* lipids is entirely consistent with homeoviscous adaptation (Nandini-Kishore *et al.*, 1977). The bacterial response to phenethyl alcohol and the response of yeasts to methanol involve changes in membrane lipids which could be consistent with a homeoviscous response, but it is by no means certain that these are, in fact, both adaptive (Nunn, 1975; Suzuki, Figami & Nakasoto, 1979). Ingram has compared a barbiturate and a phenothiazine for their ability to alter bacterial membrane lipids (Ingram *et al.*, 1978). The barbiturate, but not the phenothiazine, caused compositional changes similar to those produced by ethanol, but how these relate to tolerance or whether they represent a homeoviscous mechanism is not clear.

In conclusion, the work on bacteria and protozoans suggests that homeoviscous adaptation in cell membrane lipids occurs in response to long-chain alcohols and general anaesthetics. The position is far from clear in relation to short-chain alcohols and other drugs. At the concentrations of these other agents used in these studies it may be that considerations other than membrane fluidity influence both their toxicity and the response of the organism.

Homeoviscous adaptation in mammalian cells *in vitro*

Some of the work on homeoviscous adaptation in mammalian cells appears in a paper which has already been cited above (Ingram *et al.*, 1978). This compared the change in lipid composition induced by ethanol in bacterial cells with that produced in cultured Chinese hamster ovary (CHO) cells. The results are again difficult to interpret, partly because the cells contain very different lipids initially before exposure to ethanol. However, although one can detect similarities between the changes in lipid composition in the two cell types, the changes in the mammalian cells, which included a significant reduction in the proportion of polyunsaturated acyl chains of membrane phospholipids, are much more consistent with a homeoviscous adaptation to a fluidising agent than are the bacterial results. The concentration of ethanol used in the mammalian cell cultures was less than that in the bacterial cultures, again suggesting that the concentration of ethanol may be an important variable in determining the membrane response.

Work which indirectly suggests that mammalian cells in culture show homeoviscous adaptation to ethanol includes that by Li & Hahn (1978) where CHO cells were again used. These authors demonstrated that exposure of cultured cells to ethanol produced cross-tolerance to heat exposure. The converse was also true, cells adapted to high temperature being resistant to ethanol. Since it can be shown that thermal tolerance in CHO cells is associated with a reduced intrinsic membrane fluidity and with an increased cholesterol to phospholipid ratio (Anderson, Minton & Hahn, 1981) this suggests that a form of homeoviscous adaptation based on cholesterol incorporation is responsible for both thermal- and ethanol-tolerance in these cells.

Isolated mammalian cells have rarely been used to study homoeoviscous adaptation, although many cell types can readily be dissociated and grown in culture for extended periods.

In our laboratory we have investigated the use of human erythrocytes (J. M. Littleton, unpublished) and blood platelets (Fenn & Littleton, 1981) for studying the effect of membrane lipid composition on the cellular response to ethanol. By incubating these cells with different lipids we have found it possible to alter membrane lipid composition passively and to investigate the subsequent effect of alcohols on these transformed cells. In the case of erythrocyte haemolysis and blood platelet aggregation, the results are consistent with an alcohol-induced fluidisation of the membrane, in that alcohols produce similar effects to incorporation of unsaturated acyl chains. However, in the case of platelet aggregation the dose–response curve for inhibition by ethanol is steeper when saturated acyl chains are incorporated. Assuming that sensitivity to aggregating agents in some way directly reflects fluidity of platelet membranes, this gives rise to the paradoxical observation that, at high concentrations of alcohol, platelets with saturated acyl chains incorporated are less excitable than those where unsaturated lipids have been incorporated. As shown in Fig. 2 this leads to a situation where incorporation of either saturated or unsaturated lipids into the membrane can be an appropriate response depending on the concentration of alcohol present.

Homeoviscous drug-adaptation in complex organisms

This section will be divided into two parts. In the first, experiments which are primarily biochemical or biophysical will be reviewed, and in the second, experiments on responses of the intact organism will be considered. Thus the first part covers the possibility that cellular responses compatible with homeoviscous adaptation occur in the intact organism *in vivo*. The second part describes the extent to which ethanol and hyperthermia produce similar effects in intact animals and fish, the extent to which tolerance to one agent induces tolerance to the other, and also whether alteration *in vivo* of lipid metabolism can affect the subsequent development of ethanol tolerance.

Fig. 2. The hypothetical situation which may arise when incorporation of more fluid lipids (circles) into a cell membrane has the same effect as a 'fluidising' drug but makes the concentration–response relationship to that drug more shallow. Thus in the absence of drug the response of the cell is reduced from its optimum level by incorporation of more fluid lipids whereas the incorporation of lipids which are less fluid (squares) has the opposite effect. As concentrations of the fluidising drug increases, depressing function below the optimum level, an appropriate adaptation is to increase the proportion of lipids which are more viscous. However, at drug concentrations above that represented by the broken line the response of cells with viscous lipids incorporated into the membrane is depressed to a greater extent than that of cells with fluid lipids incorporated. An appropriate adaptive response at these concentrations is therefore to increase the proportion of fluid lipids in the cell membrane.

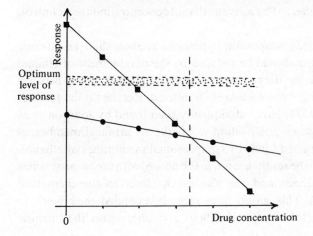

Biochemical and biophysical responses of cell membranes

Numerous pharmacological agents including alcohols, general anaesthetics and diverse other centrally active agents such as tranquillisers, barbiturates, morphine and cannabinols can increase the fluidity of the membrane lipid bilayer (Seeman, 1972). All of these compounds are capable of producing tolerance to some of their central actions, and it is theoretically possible that homeoviscous adaptation underlies at least some aspects of this tolerance. As with the previous section, most evidence is concerned with homeoviscous drug-adapation to ethanol and this will be discussed first.

Direct evidence that prolonged administration of ethanol to mammals produced a form of homeoviscous adaptation was first obtained by Chin & Goldstein (1977). Using an electron spin resonance method with mouse synaptosomal membranes they showed that the fluidisation produced by ethanol *in vitro* was reduced if the membranes were taken from mice which had been made tolerant to ethanol; a similar effect was also noted in erythrocyte membranes.

At first sight this may seem exactly what is predicted by the homeoviscous drug-adaptation hypothesis. Thus the membranes of the synapses and erythrocytes appear to have adapted to prevent or to mitigate the fluidisation produced by ethanol. However, there is one very puzzling feature. The homeoviscous drug-adaptation hypothesis predicts that, in response to a fluidising agent (e.g. ethanol) cell membranes should accumulate lipids which decrease fluidity in order to overcome the fluidisation. This should make the 'intrinsic' fluidity of these membranes less than that of control membranes when fluidity is measured in the absence of the fluidising agent. This was not the case. Chin & Goldstein found no measurable difference in the intrinsic fluidity of control and 'tolerant' membranes in the absence of ethanol. It was only when ethanol was added that the difference was expressed, tolerant membranes being less affected by a given ethanol concentration than controls (see Fig. 3).

The homeoviscous drug-adaptation hypothesis predicts that the intrinsic fluidity of the membrane should be reduced by the development of ethanol tolerance (Fig. 3*a*), and that the concentration–response relationship between ethanol and fluidity should not necessarily be altered. Results similar to those of Chin & Goldstein (1977) have subsequently been found by Johnson *et al.* (1979) using a fluorescence polarisation technique on artificial membranes prepared from the extracted lipids of synaptosomal membranes of ethanol-tolerant animals. Again the results are not in full accord with the homeoviscous drug-adaptation hypothesis and are also much closer to the theoretical treatment in Fig. 3(*b*). This group have recently extended their work to include other species (Johnson *et al.*, 1980) and once again the intrinsic fluidity of the membranes of ethanol-tolerant animals does not differ significantly from controls, but the concentration–response relationship is shallower. This, therefore, seems to be a general principle and one which only partially supports the original concept of homeoviscous adaptation as applied during temperature acclimation.

There is an important theoretical reason which could explain this discrepancy between the predicted and actual results. The homeoviscous adaptation hypothesis is based on the concept of adaptation to a constant concentration of fluidising agent in the cell membrane. In order to maintain membrane fluidity within the normal range, the cell simply has to adjust its intrinsic membrane fluidity downwards by incorporating more saturated lipids. No alteration in the concentration–response relationship is necessary because the concentration of fluidising agent is assumed to be constant. This is, however, not the case in the animal experiments cited above, where ethanol tolerance is induced by providing ethanol in a liquid diet. Under these conditions one would predict that ethanol intake would fluctuate in a diurnal pattern and

Fig. 3. The relationship between the concentration of ethanol and membrane viscosity in control membranes (triangles) and ethanol-adapted membranes (circles) in two different theoretical models. In (*a*) the ethanol-adapted membranes have a higher viscosity in the absence of drug but at a constant ethanol concentration, represented by the broken line, they have an identical viscosity to control membranes. In (*b*) there is no difference between the control and adapted membrane viscosity in the absence of ethanol but the relationship of concentration to reduction in viscosity is more shallow. Thus, although there is no concentration of ethanol at which membrane viscosity of adapted membranes is identical with that of control membranes, the adapted membranes are protected against fluctuations in ethanol concentrations more effectively than those in (*a*). The model shown in (*b*) fits the experimentally obtained results better than the homeoviscous adaptation hypothesis (*a*).

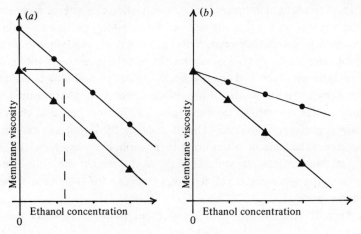

consequently blood ethanol and brain ethanol concentrations would also fluctuate. In these circumstances, it may well be more appropriate for the cell to change its membrane characteristics so that the concentration–response relationship for ethanol is less steep, since this would better protect the cell membrane from the effects on fluidity of variations in the concentration of ethanol. The situation is analogous to that in Fig. 2 where a flattening of the concentration–response relationship becomes important as the concentration rises above a certain point. The differences between these two responses may partially account for discrepancies in results from different laboratories. It would be interesting to compare results from animals in which ethanol tolerance had been induced by a technique which allowed fluctuations in concentration with results from animals in which a steady increase to a more stable concentration (e.g. inhalation: see Littleton & John (1977) below) is produced. It seems likely that the mechanisms of these two adaptations, if both exist, will prove to be different (see Fig. 2).

The biophysical measurements described above clearly suggest that some form of ethanol tolerance does exist at the level of cell membranes, and that the mechanism involves an alteration in the *lipid* composition of these membranes (the work of Johnson and others utilises only the extracted lipids of synaptosomal membranes). The biochemical nature of this change is still unclear but two main alterations have been described; an increase in saturation of phospholipid acyl chains in synaptosomal membranes (Littleton & John, 1977) and in other tissues (Littleton, John & Grieve, 1979), and secondly the increase in cholesterol to phospholipid ratio reported in synaptosomal and erythrocyte membranes (Chin, Parsons & Goldstein, 1978). It is not yet known whether either of these changes really constitutes an appropriate adaptive change to the presence of ethanol. Johnson *et al.* (1979) found that the differences in biophysical characteristics between control and tolerant membranes disappeared when cholesterol was removed from the membrane, but that the *amount* of cholesterol present was relatively unimportant. It has also been reported (Littleton *et al.*, 1980) that only certain strains of mice show a rapid development of tolerance to ethanol with an associated change in the acyl group composition of phospholipids. All strains of mice appear to show alterations in cholesterol to phospholipid ratio. Evidence so far, therefore, favours some interaction between cholesterol molecules and phospholipid acyl chains as being the important determinant of tolerance. The role of alterations in phospholipid head groups, or of binding of divalent cations such as Ca^{2+} to the membrane, should not be overlooked, though there is yet no direct evidence for their involvement in this phenomenon.

The chronic administration of drugs other than ethanol has also been studied in relation to alteration in physical and biochemical characteristics of neuronal membranes. The homeoviscous drug-adaptation hypothesis predicts that general anaesthetics, like ethanol, should also produce tolerance by reducing the intrinsic fluidity of synaptosomal membranes. Similar experiments on, and analyses of, biological membranes from animals exposed to nitrous oxide have been made, but in general the results are much less convincing than those for ethanol. Thus the synaptosomal membranes of mice made tolerant to nitrous oxide show no difference either in intrinsic fluidity, or in the concentration–response relationship of fluidisation produced by nitrous oxide, when compared with control membranes (Koblin, Dong & Eger, 1979), which also correlates with little change in lipid composition. The conclusion must be that this general anaesthetic does not elicit a homeoviscous adaptation in neurones under the experimental conditions utilised, and that the tolerance shown is due to some other adaptive process.

Curiously, there is some evidence that morphine, a drug thought to produce most of its pharmacological effects through a highly specific, receptor-mediated mechanism, may cause some homeoviscous adaptation in mammalian cells. Thus acute administration of morphine induces a membrane fluidisation, but its prolonged administration is associated with a reduction in intrinsic fluidity in both synaptosomal and erythrocyte membranes (Hosein *et al.*, 1979, 1980). The reason for this change is not yet established and it must be recognised that this, and indeed all the changes attributed to ethanol, may be coincidental and not directly related *per se* to maintenance of the fluidity of lipid bilayers.

Experiments on the responses of intact animals

If homeoviscous adaptation plays any part in the mechanism by which animals develop tolerance to general depressant drugs, then it should be possible to influence the development to one perturbation by exposure to another agent or condition which also alters membrane fluidity. We would predict, therefore, that there should be some interaction with body temperature so that tolerance to general depressants is enhanced by low body temperature (because this would reduce membrane fluidity), and also enhanced by other treatments which increase the viscosity of membrane lipids (for example increasing the proportion of saturated lipids or cholesterol). The converse predictions can also be made, that hyperthermia or increasing the proportion of unsaturated lipids, should reduce the development of tolerance.

The rapid development of central nervous system tolerance in mice has been studied by Grieve & Littleton (1979) at different ambient temperatures. At normal room temperature, ethanol causes hypothermia during the development of tolerance and this can be prevented by keeping the animals at a thermoneutral temperature. Prevention of hypothermia had no effect on the development of tolerance in two strains in which this process occurs rapidly. Mice of a strain which adapts more slowly were killed by concentrations of ethanol in blood at which mice of the same strain survived if allowed to become hypothermic. This does not argue for homeoviscous adaptation, but it does suggest similarities between the effects of ethanol and hyperthermia on the central nervous system. Results suggesting a similar conclusion have also been obtained by Malcolm & Alkana (1981) where additive effects between ethanol and hyperthermia were obtained in a variety of tests.

Evidence for cross-tolerance between hyperthermia and ethanol in peripheral tissues has recently been obtained by Anderson et al. (1982) using thermally induced tissue breakdown in the mouse ear as an end-point. These authors showed that the concomitant presence of ethanol in the bloodstream potentiates tissue breakdown but that animals made tolerant to ethanol are resistant to thermal tissue damage.

The effect on goldfish of different ambient temperatures has been used as a means of testing whether homeoviscous adaptation to n-butanol occurs (Hill, Hoyland & Bangham, 1980). The prediction is that fish at high temperatures should show an additive effect with alcohols, whereas fish in low temperatures should be protected. The results are not clear-cut: both extremes of temperature tend to sensitise the fish to alcohol when tested at an intermediate temperature. The problem, as with many of these experiments, is that the intact organism is too complex to assume that only one form of adaptation or one effect of alteration in body temperature exists.

The results of experiments in which attempts have been made to alter lipid metabolism in brain and assess changes in sensitivity to general depressants are also equivocal. Grieve et al. (1979) reported that inhibition of cholesterol synthesis in mice prevented the development of tolerance to ethanol – a result in accord with those of Chin et al. (1978) and Johnson et al. (1979) on isolated membranes. However, subsequently Ruiz & Littleton (1980) provided evidence that this finding could have been due to altered steroid hormone synthesis rather than any effect related to membrane cholesterol content. Feeding mice diets enriched in saturated lipids in one series of experiments resulted in mice which were able to develop tolerance to ethanol more rapidly than controls or animals fed unsaturated lipids (John, Littleton & Jones, 1980). In these experiments synaptosomal phospholipid acyl chain composition was not altered by diet until ethanol was given. Animals on the saturated lipid diet then incorporated more saturated acyl chains than the other groups. Similarly Koblin & Deady (1981) reported that mice fed a diet high in unsaturated lipid were affected by ethanol to a greater extent than those fed a control diet or one enriched with saturated lipids. These diets did change brain phospholipid fatty acid composition and the results, like those from our laboratory, are also in agreement with the homeoviscous drug-adaptation hypothesis. However, the authors consider that differences in body weight between the animals may have been the main reason for the altered sensitivity to ethanol. In another report (Koblin *et al.*, 1980) found no effect on nitrous oxide sensitivity of mice after dietary alteration of phospholipid fatty acid composition.

The difficulties in performance and interpretation of these experiments in intact complex organisms preclude any definitive statement on their relevance to homeoviscous drug-adaptation. Whether or not homeoviscous adaptation is important in the development of tolerance to general depressants, it seems certain that it is not the only mechanism. There may be conditions under which it is the most important mechanism but as yet these have not been established.

Conclusions

The hypothesis suggesting that organisms may adapt to drugs by a mechanism similar to that leading to temperature adaptation was a logical one. It is reasonable to suppose that organisms should utilise physiological mechanisms for adapting to pharmacological stimuli when these are appropriate. Unfortunately, the emphasis has been on homeoviscous mechanisms rather than the potentially more specific effects of lipids on, for example, membrane protein conformation, and too many assumptions have been made as to what effect pharmacological agents have on cell membranes and what might be an appropriate response to these effects. The cell membranes of the mammalian central nervous system have been perceived as relatively static, inert structures and too little attention has been given to their dynamic nature and their function. While there seems little doubt that some forms of adaptation to drugs (probably including something akin to a homeoviscous adaptation) do occur at the level of the membrane, these may well be found to be due more to changes in membrane turnover of proteins or lipids or to binding or translocation of ions than simply to the changes in lipid composition implied by the homeoviscous adaptation hypothesis.

References

Anderson, R. L., Ahier, R. G. & Littleton, J. M. (1982). Similarities between the cellular effects of ethanol and hyperthermia *in vivo*. *Radiation Research*, submitted.

Anderson, R. L., Minton, K. W., Li, G. C. & Hahn, G. (1981). Temperature-induced homeoviscous adaptation of Chinese Hamster ovary cells. *Biochimica et Biophysica Acta*, **641**, 334–8.

Cherqui, G., Cadot, M., Senault, C. & Portet, R. (1979). The lipid composition of plasma membrane and mitochondrial fractions from epididymal adipocytes of cold-acclimated rats. *Biochimica et Biophysica Acta*, **551**, 304–14.

Chin, J. H. & Goldstein, D. B. (1977). Drug tolerance in biomembranes: a spin label study of the effects of ethanol. *Science*, **196**, 684–5.

Chin, J. H. & Goldstein, D. B. (1981). Membrane-disordering action of ethanol. Variation with membrane cholesterol content and depth of the spin label probe. *Molecular Pharmacology*, **19**, 425–31.

Chin, J. H., Parsons, L. M. & Goldstein, D. B. (1978). Increased cholesterol content of erythrocyte and brain membranes in ethanol-tolerant mice. *Biochimica et Biophysica Acta*, **513**, 358–63.

Fenn, C. G. & Littleton, J. M. (1981). Inhibition of platelet aggregation by ethanol. Role of plasma and platelet membrane lipids. *British Journal of Pharmacology*, **73**, 305P–306P.

Goldman, S. S. (1975). Cold resistance of the brain during hibernation. III. Evidence of a lipid adaptation. *American Journal of Physiology*, **228**, 834–44.

Grieve, S. J. & Littleton, J. M. (1979). Ambient temperature and the development of functional tolerance to ethanol by mice. *Journal of Pharmacy and Pharmacology*, **31**, 707–8.

Grieve, S. J., Littleton, J. M., Jones, P. A. & John, G. R. (1979). Functional tolerance to ethanol in mice; relationship to lipid metabolism. *Journal of Pharmacy and Pharmacology*, **31**, 737–42.

Henle, K. J. & Dethlefsen, L. A. (1978). Heat fraction and thermotolerance. A review. *Cancer Research*, **38**, 1843–51.

Hill, M. W. & Bangham, A. D. (1975). General depressant drug dependency: a biophysical hypothesis. *Advances in Experimental Medicine and Biology*, **59**, 1–9.

Hill, M. W., Hoyland, J. & Bangham, A. D. (1980). Effects of temperature and *n*-butanol (a model anaesthetic) on a behavioural function of goldfish (*Carassius auratus*). *Journal of Comparative Physiology*, **135A**, 327–32.

Hosein, E. A., Lapalme, M., Lau, A., Wan, O., Stefansyn, H. & Zucker, J. (1980). Biphasic activity of membrane-bound enzymes in brain mitochondria and synaptosomes during the development of tolerance to, and physical dependence on, chronic morphine administration to rats. *Canadian Journal of Physiology and Pharmacology*, **58**, 484–92.

Hosein, E. A., Lapalme, M., Sacks, B. & Wiseman-Distler, M. (1979). Biphasic changes in rat brain mitochondrial membrane structure and enzyme activity after acute opiate administration to rats. *Biochemical Pharmacology*, **28**, 7–14.

Ingram, L. O. (1976). Adaptation of membrane lipids to alcohols. *Journal of Bacteriology*, **125**, 670–8.

Ingram, L. O., Buttke, T. M. & Dickens, B. F. (1980). Reversible effects of ethanol on the lipids of *E. coli*. *Advances in Experimental Medicine and Biology*, **126**, 299–338.

Ingram, L. O., Ley, K. D. & Hoffmann, E. M. (1978). Drug-induced changes in lipid composition of *E. coli* and of mammalian cells in culture – ethanol, pentobarbital and chlorpromazine. *Life Sciences*, **22**, 489–93.

John, G. R., Littleton, J. M. & Jones, P. A. (1980). Membrane lipids and tolerance to ethanol in the mouse. Effect of dietary lipid. *Life Sciences*, **27**, 545–55.

Johnson, D. A., Friedman, H. J., Cooke, R. & Lee, N. M. (1980). Adaptation of brain lipid bilayers to ethanol-induced fluidisation. Species and strain generality. *Biochemical Pharmacology*, **29**, 1673–6.

Johnson, D. A., Lee, N. M., Cooke, R. & Loh, H. H. (1979). Ethanol-induced fluidisation of brain lipid bilayers: required presence of cholesterol in membranes for the expression of tolerance. *Molecular Pharmacology*, **15**, 739–46.

Koblin, D. D. & Deady, J. E. (1981). Sensitivity to alcohol in mice with an altered brain fatty acid composition. *Life Sciences*, **28**, 1889–96.

Koblin, D. D., Dong, D. E., Deady, J. E. & Eger, E. I. (1980). Alteration of synaptic membrane fatty acid composition and anaesthetic requirement. *Journal of Pharmacology and Experimental Therapy*, **212**, 546–52.

Koblin, D. D., Dong, D. E. & Eger, E. I. (1979). Tolerance of mice to nitrous oxide. *Journal of Pharmacology and Experimental Therapy*, **211**, 317–25.

Lenaz, G., Curatola, G., Mazzanti, L., Bertoli, E. & Pastuszko, A. (1979). Spin label studies on the effect of anaesthetics in synaptic membranes. *Journal of Neurochemistry*, **32**, 1689–95.

Li, G. C. & Hahn, G. M. (1978). Ethanol-induced tolerance to heat and to adriamycin. *Nature, London*, **274**, 699–701.

Littleton, J. M., Grieve, S. J., Griffiths, P. J. & John, G. R. (1980). Ethanol-induced alteration in membrane phospholipid composition: possible relationship to development of cellular tolerance to ethanol. *Advances in Experimental Medicine and Biology*, **126**, 7–20.

Littleton, J. M. & John, G. R. (1977). Synaptosomal membrane lipids of mice during continuous exposure to ethanol. *Journal of Pharmacy and Pharmacology*, **29**, 579–80.

Littleton, J. M., John, G. R. & Grieve, S. J. (1979). Alterations in phospholipid composition in ethanol tolerance and dependence. *Alcoholism: Clinical and Experimental Research*, **3**, 50–6.

Malcolm, R. D. & Alkana, R. L. (1981). Temperature dependence of ethanol depression in mice. *Journal of Pharmacology and Experimental Therapy*, **217**, 770–5.

Nandini-Kishore, S. G., Kitajima, Y. & Thompson, G. A., Jr (1977). Membrane fluidising effects of the general anaesthetic methoxylflurane elicit an acclimation response in *Tetrahymena*. *Biochimica et Biophysica Acta*, **471**, 157–61.

Nandini-Kishore, S. G., Mattox, S. M., Martin, C. E. & Thompson, G. A., Jr (1979). Membrane changes during growth of *Tetrahymena* in the presence of ethanol. *Biochimica et Biophysica Acta*, **441**, 315–27.

Nunn, W. D. (1975). The inhibition of phospholipid synthesis in *E. coli* by phenethyl alcohol. *Biochimica et Biophysica Acta*, **380**, 403–13.

Ruiz, J. S. & Littleton, J. M. (1980). Glucocorticoids play an important role in the rapid development of tolerance to ethanol by mice. *Substance/Alcohol Actions/Misuse*, **1**, 415–22.

Seeman, P. (1972). The membrane actions of anaesthetics and tranquillizers. *Pharmacological Reviews*, **24**, 583–655.

Singer, S. J. (1981). Current concepts of molecular organisation in cell membranes. *Biochemical Society Transactions*, **9**, 203–6.

Suzuki, O., Figami, Y. & Nakasoto, S. (1979). Changes in lipid composition of methanol-grown *Candida guillermondi*. *Agricultural and Biological Chemistry*, **43**, 1343–6.

H. O. J. COLLIER

Cellular adaptation of receptor-mediated function during opiate exposure

Adaptation to opiates

Our main knowledge of opioid action and adaptation to it derives from experiments performed on mammals. The response of a mammal to continued treatment with opiate takes two forms: tolerance and dependence. Tolerance, which is characterised by a diminished response to drug as exposure continues, has an obvious adaptive quality. Dependence, which normally accompanies tolerance to opiates, is usually characterised by (i) a heightened response to specific opiate antagonists, and (ii) a behavioural disturbance on drug withdrawal (the withdrawal syndrome). A third possible characteristic of dependence – an enhanced excitability on drug withdrawal – should also be noted, since, alone of these characteristics, it applies to the mouse vas deferens (see below).

Many years ago I described dependence as the price paid for the adaptive process of tolerance (Collier, 1966). Today, however, the discovery of the endogenous opioids, which in high doses induce dependence (Wei & Loh, 1976; Bläsig & Herz, 1976), has led to the suggestion that the withdrawal syndrome might have an adaptive function, perhaps as a rapid means of throwing off the profound inhibitory effect of continued exposure to opioid, when this ceases. Such a concept is exemplified by the proposal of Margules (1979) that endogenous opioids act as mediators of a state of torpor or hibernation in times of reduced food supply, and that the withdrawal syndrome provides a rapid arousal mechanism when the need for torpor has passed. He postulated also the existence of an endogenous opioid antagonist, termed 'endoloxone'; but this does not appear to be essential to the concept of withdrawal behaviour as an arousal mechanism.

I propose, therefore, to consider both tolerance and dependence as adaptive responses to continued treatment with opiate and to turn to the uncertainty principle that applies to their relationship. Since we cannot measure tolerance without administering drug, nor dependence without withdrawing it, we cannot measure both tolerance and dependence in the same preparation at the same time (Collier, 1973).

If tolerance and dependence are measured at the same time in different animals, exposed in parallel to opiate, a much greater degree of dependence than of tolerance is produced (Way, Loh & Shen, 1969). The principle of uncertainty makes it hard to interpret this difference. For this reason, I propose to refer to the response to continued exposure to opiate as 'opiate dependence and associated tolerance' (ODT).

Receptor-mediated function in neurones

In man and other mammals, morphine has a variety of actions, which Claude Bernard succinctly described as a mixture or a succession of the depressant and the stimulant. When we focus down on the production of nerve impulses by individual neurones *in situ* in the brains of experimental animals, we find that iontophoretic application of opioid may stimulate or inhibit a neurone, according to its type. Both effects are blocked by low concentrations of the specific opioid antagonist naloxone, indicating that these are specific actions mediated through opioid receptors.

For example, Nicoll *et al.* (1977), in a survey of individual neurones *in situ* in different regions of rat brain, found that normorphine, β-endorphin or met-enkephalin inhibited nerve impulse production in the majority of responsive cells in the brainstem, cerebral cortex, caudate nucleus and thalamus, whereas each opioid excited the hippocampal pyramidal cells. A subsequent analysis showed, however, that in the hippocampus the excitation of pyramidal cells could be attributed to an inhibition by opioid of the closely adjoining basket cells, which themselves exert an inhibitory control on the pyramidal cells (Zieglgänsberger *et al.*, 1979; Nicoll, Alger & Jahr, 1980). In short, an apparent excitatory effect of opioids may be a disinhibition.

Not all the excitatory actions of opioids have yet been explained as inhibitions of inhibition. For example, the excitatory action of morphine on the circular muscle of rat colon, recently studied by Huidobro-Toro & Way (1981), has not been so explained. The findings, however, that (i) the excitatory action of morphine in the colon is mediated through an excitatory serotonergic neurone, and (ii) that, in tolerant and dependent preparations, tolerance is expressed as diminished excitation by morphine whereas dependence is expressed as increased excitation on withdrawal, strongly suggest that opiate here acts by inhibiting a neurone that itself inhibits an excitatory serotonergic neurone supplying the circular muscle.

These considerations, coupled with the finding that selective activation of each of the three varieties of opiate receptor so far distinguished (μ, δ and κ) leads to inhibition of *in vitro* preparations, indicate that opiates exert a primarily inhibitory effect.

Opiate dependence and tolerance *in vitro*

Opiate dependence and associated tolerance were first observed in man, as an unwanted effect of continued use of opiates. Later, it was found that a comparable condition of ODT could be reproduced in laboratory animals, first in dogs and monkeys and then in rats, mice and guinea-pigs. It is only quite recently, however, that it has been established that ODT, resembling in essential characteristics that observed in whole mammals, can be reproduced in neurones *in vitro* and in isolated mammalian cells of neural origin in culture. This discovery shows that ODT is not, or need not be, an integrated response of assemblages of neurones, but is a cellular phenomenon, developing within cells that bear opiate receptors (North & Karras, 1978; Collier, 1978, 1980). That ODT can be produced in defined cells *in vitro* provides a practical means of analysing its mechanism.

So far, ODT has been induced *in vitro* with a reasonable degree of certainty in two types of cell, both characterised by the possession of opiate receptors. One is a normal neurone, the final cholinergic motoneurone (FCMN) of the myenteric plexus, supplying the longitudinal smooth muscle of guinea-pig ileum. The opiate receptors of this neurone were investigated by Kosterlitz & Watt (1968), and, more recently, they have been found to belong largely to the μ subtype of opiate receptor, which is particularly responsive to morphine and allied opiates (Lord *et al.*, 1977). The second type of cell in which ODT has been obtained *in vitro* is a line of neuroblastoma × glioma hybrid cells maintained in culture, usually designated NG 108-15 cells. Opiate receptors were first detected on cells of this line by Klee & Nirenberg (1974). Let us consider ODT in each of these two types of cell in turn.

ODT in guinea-pig ileum

When the guinea-pig isolated ileum, suspended in Krebs solution, is exposed to repeated electrical field stimulation adjusted to excite the FCMN of the myenteric plexus, the longitudinal smooth muscle of the ileum contracts in response to the release of acetylcholine at the terminal after each stimulus. Opiates inhibit this response in a dose-dependent manner (Paton, 1957; Schaumann, 1957). If exposure to opiate continues, the inhibitory effect readily develops tolerance (Paton, 1957; Fennessy, Heimans & Rand, 1969). If the preparation, after sufficiently prolonged exposure to morphine, is challenged with naloxone, an opioid antagonist, the longitudinal muscle contracts sharply in the absence of electrical stimulation in response to release of acetylcholine from the terminal of the FCMN elicited by naloxone (Ehrenpreis, Light & Schonbuch, 1972). This contracture to naloxone is not seen after brief exposure of the preparation to morphine.

Fig. 1 illustrates a convenient method of inducing and measuring ODT in the isolated ileum (Hammond, Schneider & Collier, 1976; Collier, Cuthbert & Francis, 1981b). We have used this method to answer the question of how far the properties of the ileum, after incubation with opiate, resemble dependence in the whole animal. Table 1 summarises these properties, which are precisely the same as those generally regarded as characteristic of ODT in the whole animal. Hence there seems little doubt but that opiate dependence with associated tolerance can be induced in the guinea-pig ileum *in vitro*. The main difference between the ileum and the whole animal is that whereas in the ileum only one withdrawal effect, contracture of the longitudinal muscle, can readily be observed, in the whole animal many different behavioural effects express the withdrawal disturbance. Another difference, that dependence in the ileum reaches a peak in about 6 hours whereas it takes a day or more to develop in the whole rodent, is probably more apparent than real, since, when morphine was continuously infused into a vein of the

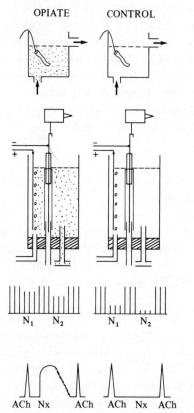

Fig. 1. Induction and expression of opiate tolerance and dependence in guinea-pig ileum *in vitro*.

Pieces from the same ileum placed in Krebs solution containing hexamethonium with or without opiate and/or other test drug

Incubated at 5 °C or 22 °C for 18–24 hours or at 37 °C for 8 hours

Set up in incubation fluid at 37 °C in parallel organ baths for recording responses to added drugs or electrical stimuli

Tolerance: Shift to right of D/R line for inhibition of electrically evoked contracture of ileum. N_1, lower, and N_2, higher dose of normorphine

Dependence: Dose-related contracture of opiate-incubated preparations to naloxone (Nx), compared with standard dose of acetylcholine (ACh)

rat, a peak of tolerance was observed within a few hours (Cox, Ginsburg & Osman, 1968). The rate of induction probably therefore relates to the opiate concentration continuously applied to the cells that will develop ODT.

Cellular site of ODT in the ileum

The question arises as to which particular cell, if any, among the many types present in the ileum, including various mucosal, smooth muscle and neural cells, dependence arises in. Pharmacological evidence points to the FCMN as the site of the opiate dependence in the isolated ileum described above. This evidence is three-fold.

First, there is the evidence that opiates and naloxone act acutely on the myenteric plexus (Paton & Zar, 1968); and, since hexamethonium, an acetylcholine/nicotinic receptor antagonist, does not interfere with opiate action (Greenberg, Kosterlitz & Waterfield, 1970), probably on the FCMN. This evidence is supported by studies of the acute action of opiates on nerve impulse production by myenteric neurones when transmission between neurones is blocked (Dingledine & Goldstein, 1976; North & Tonini, 1977).

Secondly, ODT can be induced in the ileum in the presence of hexamethonium in sufficient concentration to block descending impulses from the cholinergic neurones supplying the FCMN (Hammond, Schneider & Collier, 1976; North & Karras, 1978; Collier *et al.*, 1981*b*). Moreover, the degree of ODT is comparable when induced in the presence or absence of hexamethonium (Hammond *et al.*, 1976; H. O. J. Collier, N. J. Cuthbert & D. L. Francis, unpublished).

Thirdly, in dependent preparations, atropine or hyoscine, which prevent

Table 1. *Characteristics of opiate dependence found in both guinea-pig isolated ileum and whole animals*

1 Dependence is accompanied by tolerance to the acute effect of opioids.
2 Dependence does not appear immediately, but develops on continued exposure to opiate.
3 Withdrawal of opiate elicits a behavioural disturbance that can be suppressed by retreatment with opiate.
4 A withdrawal disturbance can be precipitated with a specific opiate antagonist, the potency of which is inversely related to the intensity of dependence.
5 Induction of dependence requires the activation of a specific and stereospecific opiate receptor.
6 Precipitation of withdrawal requires a specific and stereospecific opiate antagonist.
7 Clonidine inhibits the opiate withdrawal disturbance.

From Collier *et al.* (1981*a, b*).

acetylcholine released at the terminal of the FCMN from exciting the longitudinal smooth muscle, inhibits the contractile response to naloxone challenge (Ehrenpreis *et al.*, 1972; Schulz & Herz, 1976; North & Karras, 1978).

The evidence that ODT occurs solely in the FCMN, without an essential contribution from other neurones is not, however, complete, because the FCMN probably receives other excitatory and inhibitory inputs. Among the excitatory inputs that the FCMN may also receive are those from cells which liberate 5-hydroxytryptamine, substance P, prostaglandin E_2 and cholecystokinin (see review by Gabella, 1981). Among inhibitory inputs that the FCMN may receive are those from cells producing noradrenaline, adenosine* or other purinergic transmitter, gamma-aminobutyric acid and somatostatin. Even though specific blocking drugs are available for several of the receptors involved, other receptors cannot yet be specifically blocked and other mediators not tested above may also interact with the FCMN. Hence, it is not yet possible to conclude that dependence within the FCMN does not require input from some other neurone or neurones, although the evidence given in the papers cited above makes it very probable that opiate dependence does develop in that neurone.

Site of ODT within the FCMN

Fig. 2 illustrates diagrammatically some of the receptors that the FCMN probably possesses and indicates the different functional parts in which the changes producing ODT might develop. These parts are four: receptors, soma, axon and terminal. Let us consider each in turn as a possible site of the mechanism of ODT.

Receptors. The term 'receptor' has various usages, but, in this discussion, it is used for a functional complex in or on the cell which not only recognises and binds a foreign agonist molecule but also makes an initial response to it. A receptor is therefore thought to consist of a specific recognition site that binds only a particular type of ligand. Since specific antagonists are also thought to bind with this site, without producing any reaction beyond blocking the binding of an agonist molecule, an effector part of the receptor complex is postulated, which responds to agonist but not to antagonist. To link the recognition site and the effector, a coupling unit may also be postulated (Fig. 3).

As Table 2 indicates, one possible mechanism of adaptation to opiate or other drug is that continued activation by a drug of its specific receptors would

*Note added in proof: We have recently shown that adenosine induces a dependence in the FCMN resembling that of opiates or clonidine, but mediated through a distinct inhibitory receptor (H. O. J. Collier & J. F. Tucker, unpublished).

lead to a reduction in their density at the site receiving opiate (Axelrod, 1956; Collier, 1966). This would be expected to lead to tolerance, since the probability of an interaction between opiate molecule and its binding site would be reduced. It would not, however, in itself seem likely to lead to dependence, in which the dependent neurone develops increased endogenous excitation.

Most attempts to show a reduction of specific opiate binding sites in brain or ileum preparations from animals made tolerant to morphine have failed (Lee *et al.*, 1973; Hitzemann, Hitzemann & Loh, 1974; Klee & Streaty, 1974; Pert & Snyder, 1976; Cox & Padhya, 1977; Hollt & Wüster, 1978). Davis, Akera & Brody (1979) have, however, found a reduction in opiate binding sites in brainstem slices of the rat, accompanying continued treatment with morphine and associated with morphine tolerance. The old proposal of a

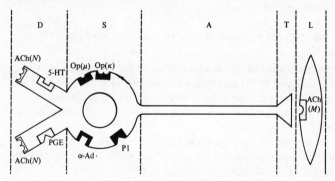

Fig. 2. Functional divisions of the final cholinergic motoneurone of the myenteric plexus supplying the longitudinal smooth muscle of the guinea-pig ileum and some receptor types probably represented thereon. D, dendrites; S, soma with nucleus; A, axon; T, terminal; L, longitudinal smooth muscle. Receptors that lead to excitation when activated are shown as white: ACh(N), nicotinic (postganglionic) acetylcholine; 5-HT, 5-hydroxytryptamine; PGE, E prostaglandin; ACh(M), muscarinic acetylcholine. Receptors that lead to inhibition when activated are shown as black: Op(μ), opiate of μ subtype; Op(κ), opiate of κ subtype; α-Ad, α-adrenoceptor; P_1, purinergic of type P_1.

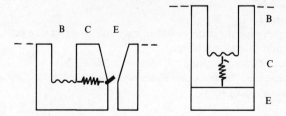

Fig. 3. Possible arrangements of receptor complexes. B, binding or recognition site; C, coupling unit; E, effector (ion channel, adenylate cyclase, etc.).

reduction in density of opiate receptors as a mechanism of opiate tolerance cannot therefore be entirely abandoned.

Some years ago I proposed that the excitation that characterises withdrawal from opiate of a dependent preparation, might be due to induction by opiate treatment of receptors for an excitatory transmitter (Table 2). Although I did not at that time use a term for this postulated process, it now seems to me that the term 'contra-induction of receptors' would be suitable. By this I mean the induction, by administration of one substance having receptors on a cell, of specific receptors on the same or an adjoining cell for another substance having the opposite pharmacological effect.

A few instances of contra-induction of receptors have been described in the recent literature (Llorens et al., 1978; Schulz, Wüster & Herz, 1979; Dum et al., 1979; Syapin & Rickman, 1981), but for two reasons it seems unlikely that contra-induction of excitatory receptors in response to the inhibitory action of opiate explains the major part of ODT in the FCMN. First, the ileum removed from a guinea-pig made intensely tolerant to and dependent on morphine is supersensitive not only to several excitatory substances acting through different receptors – 5-hydroxytryptamine, E prostaglandins and nicotine – but also to electrical or potassium-induced stimulation (Schulz & Herz, 1976; Cox & Padhya, 1977; Johnson et al., 1978). It would be difficult to explain such a wide range of increased sensitivities, particularly that to electrical stimulation, in terms of contra-induction of receptors for one excitatory transmitter.

Table 2. *Summary of receptor theory of tolerance and dependence*

1. Drug treatment may bring about a change in the density or efficiency of receptors for drug or for endogenous substance with which it interacts.
2. Changes in density or efficiency of receptors may be adaptive.
3. Tolerance could arise from a reduction in density or efficiency of drug receptors.
4. Tolerance and dependence could arise from:
    (i) an induction of receptors for an endogenous substance having an action opposed to the drug's (contra-induction);
    (ii) an increased efficiency of such receptors;
    (iii) a reduction in density of receptors for an endogenous substance synergising with the drug;
    (iv) a reduced efficiency of such receptors.
5. Sensitisation could arise from:
    (i) induction of receptors for the drug or for an endogenous substance acting in the same direction (mimeo-induction);
    (ii) increased efficiency of such receptors.

Based on Collier (1965, 1966, 1969, 1973).

A second reason that such a receptor mechanism is not likely to be primarily involved in opiate dependence of the ileum stems from our observations that clonidine dependence can be induced by incubation of an isolated ileum with clonidine (an α-adrenoceptor agonist), and withdrawal elicited by challenge with the α-adrenoceptor antagonists, phentolamine or yohimbine (Collier, Cuthbert & Francis, 1981a). In clonidine-dependent ileum naloxone will not elicit a contracture, but opiate will suppress the contracture elicited by phentolamine. Likewise, in opiate-dependent ileum phentolamine will not elicit a contracture, but clonidine will suppress the contracture elicited by naloxone. Since both clonidine and opiate inhibit impulse production by the FCMN (North & Tonini, 1977; Morita & North, 1981; Tokimasa, Morita & North, 1981), it is reasonable to suppose that dependence on both inhibitory substances occurs at a post-recognition site level in the neurone.

Axon. There is no evidence that, in the acute action of morphine or in morphine dependence, the nerve impulse conducted in the axon is affected in size. However, the frequency of impulse production, arising from the soma near the axon, is greatly increased during withdrawal (Karras & North, 1981).

Terminal. An early theory of the mechanism of opiate tolerance and dependence proposed that release of acetylcholine was inhibited by the continued action of morphine at the nerve terminal (Paton, 1969). This inhibition was supposed to form a reservoir of unreleased acetylcholine. As the pressure built up in this reservoir, leakage of acetylcholine would produce tolerance and, when morphine was withdrawn, a sudden release of the accumulated acetylcholine would generate the withdrawal effects. Although evidence for this ingenious theory has been sought, no convincing evidence has yet been reported.

Some positive evidence against the terminal of the FCMN being the site of ODT has, on the contrary, been obtained. It is that tetrodotoxin, which blocks axonal conduction, prevents the expression of a withdrawal contracture of the guinea-pig ileum (Schulz & Herz, 1976; Collier et al., 1981b). If the terminal were the site of dependence, tetrodotoxin would not be expected completely to block the withdrawal effect. This observation suggests that dependence in the FCMN arises in a pre-axonal site.

Soma. The foregoing considerations point to the soma of the FCMN as the probable site of ODT. This conclusion is consistent with the electrical studies of North & Karras (1978) and Karras & North (1981) on nerve impulse production by somata of dependent neurones of the myenteric plexus. In these studies a suction electrode was attached to the soma of a neurone of the plexus to enable nerve impulse production from single neurones to be recorded. Drugs were administered by addition to the shallow organ bath in which the neurones, attached to longitudinal muscle, were observed. Neurones were utilised which in these conditions spontaneously produced nerve impulses over long periods.

Neurones derived from guinea-pigs treated with high doses of morphine for 2 or 3 days, or incubated with opiate *in vitro* for 24 hours at 22–26 °C, were less responsive to inhibition by normorphine than were controls. Tolerant neurones also responded vigorously to challenge with naloxone, by generating impulses at rates of 10–20 Hz, which is higher than rates seen in control neurones of this plexus (Karras & North, 1981). Thus, dependence and associated tolerance develop in the soma of myenteric neurones through changes that lead to or permit exceptionally high rates of generation of nerve impulses. We do not yet know what these changes are, but the mechanism is open to pharmacological analysis, using either impulse production by the soma or acetylcholine release by the terminal, or both, as measures of effect.

ODT in neuroblastoma × glioma cells

Although the mechanism of ODT in the soma of the FCMN is poorly understood, some pioneering work has elucidated the molecular mechanism of dependence and associated tolerance in cultures of a mouse neuroblastoma × rat glioma hybrid cell line, strain NG 108-15, and points the way to experiments on the soma of the FCMN.

Although the NG 108-15 cells look somewhat like neurones, have long processes and can even produce action currents on their surfaces, they differ from the neurones of the myenteric plexus in that they do not have an organised relationship between each other. To the best of our knowledge, each cell is like its neighbours, although the cells may differ in age, since they divide from time to time.

These cells possess opiate receptors that fulfil the usual requirements of specificity and stereospecificity (Klee & Nirenberg, 1974). The immediate result of interaction of the opioid molecule with its receptor is the inhibition of adenylate cyclase, with consequent lowering of the cellular level of cyclic AMP (Traber *et al.*, 1975; Sharma, Nirenberg & Klee, 1975*b*). If exposure to opioid continues, adenylate cyclase activity increases, to generate more cyclic AMP in compensation. This intensified enzyme activity produces tolerance and, if opiate is withdrawn by giving the cells naloxone, a burst of cyclic AMP production results (Sharma, Klee & Nirenberg, 1975*a*, 1977; Lampert, Nirenberg & Klee, 1976). Fig. 4 summarises this mechanism.

We do not yet know whether a compensating hypertrophy of the cyclic

AMP system (increase in effective concentration of adenylate cyclase by whatever means) is the mechanism of ODT in normal neurones, although it was originally proposed as such (Collier & Roy, 1974) and there is a good deal of indirect evidence that it is (Collier & Francis, 1975; Collier, 1980). It would therefore be desirable to test whether opiates inhibit adenylate cyclase in the soma of the FCMN and, if so, whether a compensatory hypertrophy of the enzyme occurs. It would also be necessary to determine the physiological effects of cyclic AMP in the soma of the FCMN. These things are not easily determined, because adenosine interacts with an inhibitory purinergic receptor on the myenteric plexus (Griffith et al., 1981).

Fig. 4. Model of the role of adenylate cyclase regulation in the development of morphine tolerance and dependence in neuroblastoma × glioma hybrid cells in culture. (a) The effects of morphine upon cyclic AMP levels. (b) The effects of the opiate on adenylate cyclase activity as a function of time. (From Sharma et al. (1975a), with permission.)

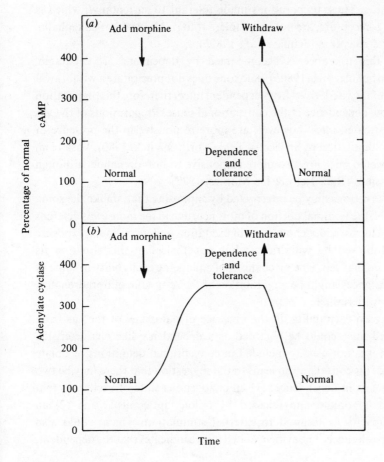

Mouse vas deferens

The mouse vas deferens provides another system that is sensitive to opioids *in vitro* (Henderson, Hughes & Kosterlitz, 1972; Hughes, Kosterlitz & Leslie, 1975). In this preparation, only the axon and terminal portion of the adrenergic motoneurone remain, the soma and its adjoining structures having been left behind during excision of the vas. When the vas deferens is subjected to electrical field-stimulation, the motoneurone releases noradrenaline, which contracts the smooth muscle. Opioids inhibit this noradrenaline release in a specific, stereospecific and dose-related way, indicating that the fragment of axon and terminal of the motoneurone in the vas possesses opiate receptors, whose activation leads to the inhibition of transmitter release.

When this preparation is derived from mice that have received continued intense treatment with opioid, it exhibits tolerance (Schulz *et al.*, 1980*a*, *b*; North & Vitek, 1980). This tolerance is highly selective for each variety of opiate receptor. Thus, preparations made tolerant to surfentanyl, which is selective for μ-receptors, are not cross-tolerant to D-Ala^2-D-Leu^5-enkephalin, selective for δ-receptors (Schulz *et al.*, 1980*b*).

Whether this tolerance is accompanied by dependence is debatable. According to Schulz *et al.* (1980*b*), naloxone does not precipitate a withdrawal contracture of the vas derived from dependent mice; therefore this preparation does not show dependence in the conventional sense. Preparations of the vas from opiate-treated mice, however, are more responsive in the presence of naloxone to stimulation by an electric field (North & Vitek, 1980). Vasa from morphine-treated mice are also more responsive to noradrenaline, although not to potassium (Rae, Neto & De Moraes, 1977).

These observations may be interpreted by supposing that, unlike the soma of the FCMN, the terminal portion of motoneurone in the mouse vas deferens does not develop an endogenous state of excitation under continued exposure to opioid, followed by withdrawal; but nonetheless in this situation its excitability is enhanced. One mechanism of enhancing excitability in some of these circumstances might be an inhibition of the re-uptake of noradrenaline from the nerve terminal.

This increased excitability, in the presence of naloxone, of the vas from opiate-treated mice could be regarded as a dependence-like characteristic, even though it is not usually included in conventional definitions of opiate dependence. This consideration leads to the suggestion that there may be two distinct elements in dependence: (1) an endogenous state of excitation that is restrained by opioid and released by opioid antagonists; and (2) an increased capacity to respond to external stimulation, which would also contribute to tolerance. The evidence cited above indicates that the dependent, withdrawn FCMN of the guinea-pig ileum exhibits both of these properties, whereas the terminal of the adrenergic motoneurone of the mouse vas deferens only exhibits the latter. The mouse vas deferens therefore also provides a preparation for the analysis of the mechanisms of tolerance and dependence.

Summary and conclusions

(1) Continued exposure to opiate of neurones possessing opiate receptors induces tolerance and dependence, both of which can be regarded as adaptive, since they represent an increased excitation or excitability, counteracting the primarily inhibitory action of the opiate.

(2) Tolerance cannot be measured without administering drug, nor dependence without withdrawing it. Hence it is impossible to measure tolerance and dependence in the same preparation at the same time. This principle of uncertainty makes it convenient to speak of opiate dependence and associated tolerance (ODT).

(3) Mechanisms of ODT cannot easily be analysed *in vivo*; but, in (i) the final cholinergic motoneurone (FCMN) of the myenteric plexus of guinea-pig ileum and (ii) the mouse neuroblastoma × rat glioma hybrid cell, strain NG 108-15, ODT can be induced and analysed *in vitro*.

(4) Pharmacological and electrophysiological analyses show that ODT develops in the soma of the FCMN, after opiate binding to the recognition site and before transmission of the nerve impulse down the axon.

(5) There is direct evidence that the molecular mechanism of ODT in NG 108-15 cells is an increased activity (hypertrophy) of adenylate cyclase in response to inhibition by opioid. There is indirect evidence that this mechanism applies also to ODT in intact animals. Evidence on the molecular mechanism of ODT in the soma of the FCMN remains to be obtained.

(6) Comparisons of the responses to naloxone and electrical stimulation of the vasa deferentia of mice that have been subjected or not subjected to continued treatment with opioid suggest that there are two distinct elements in cellular dependence: (i) an endogenous state of excitation that is restrained by opioid and released by opioid antagonist, and (ii) an increased responsiveness to external stimulation, which would also contribute to tolerance.

(7) Whereas the FCMNs of the myenteric plexus of guinea-pig ileum and NG 108-15 cells exhibit both elements of dependence, the terminal portion of the adrenergic motoneurone of the mouse vas deferens, after exposure to morphine *in vivo*, has so far been shown only to exhibit an increased capacity to respond to external stimulation. This preparation therefore provides a further model for the analysis of ODT.

References

Axelrod, J. (1956). Possible mechanism of tolerance to narcotic drugs. *Science*, **124**, 263–4.

Bläsig, J. & Herz, A. (1976). Tolerance and dependence induced by morphine-like pituitary peptides in rats. *Naunyn-Schmiedeberg's Archives of Pharmacology*, **294**, 297–300.

Collier, H. O. J. (1965). A general theory of the genesis of drug dependence by induction of receptors. *Nature, London*, **205**, 181–2.

Collier, H. O. J. (1966). Tolerance, physical dependence and receptors: a theory of the genesis of tolerance and physical dependence through drug-induced changes in the number of receptors. *Advances in Drug Research*, **3**, 171–88.

Collier, H. O. J. (1969). Humoral transmitters, supersensitivity, receptors and dependence. In *Scientific Basis of Drug Dependence (a Biological Council Symposium)*, ed. H. Steinberg, pp. 49–66. London: J. & A. Churchill.

Collier, H. O. J. (1973). Pharmacological mechanisms of drug dependence. In *Pharmacology and the Future of Man, Proceedings of the Fifth International Congress of Pharmacology*, vol. 1, ed. J. Cochin & E. L. Way, pp. 65–76. Basel: Karger.

Collier, H. O. J. (1978). Biochemical theories of opioid dependence: an analysis. In *Proceedings of the European Society for Neurochemistry, Second Meeting in Göttingen*, vol. 1, ed. V. Neuhoff, pp. 374–85. Weinheim & New York: Verlag Chemie.

Collier, H. O. J. (1980). Cellular site of opiate dependence. *Nature, London*, **283**, 625–9.

Collier, H. O. J., Cuthbert, N. J. & Francis, D. L. (1981a). Clonidine dependence in the guinea-pig isolated ileum. *British Journal of Pharmacology*, **73**, 443–53.

Collier, H. O. J., Cuthbert, N. J. & Francis, D. L. (1981b). Model of opiate dependence in the guinea-pig isolated ileum. *British Journal of Pharmacology*, **73**, 921–32.

Collier, H. O. J. & Francis, D. L. (1975). Morphine abstinence is associated with increased brain cyclic AMP. *Nature, London*, **255**, 159–62.

Collier, H. O. J. & Roy, A. C. (1974). Morphine-like drugs inhibit the stimulation by E prostaglandins of cyclic AMP formation by rat brain homogenate. *Nature, London*, **248**, 24–7.

Cox, B. M., Ginsburg, M. & Osman, O. H. (1968). Acute tolerance to narcotic analgesic drugs in rats. *British Journal of Pharmacology and Chemotherapy*, **33**, 245–56.

Cox, B. M. & Padhya, R. (1977). Opiate binding and effect in ileum preparations from normal and morphine pretreated guinea-pigs. *British Journal of Pharmacology*, **61**, 271–8.

Davis, M. E., Akera, T. & Brody, T. M. (1979). Reduction of opiate binding to brainstem slices associated with the development of tolerance to morphine in rats. *Journal of Pharmacology and Experimental Therapeutics*, **211**, 112–19.

Dingledine, R. & Goldstein, A. (1976). Effect of synaptic transmission blockade on morphine action in the guinea-pig myenteric plexus. *Journal of Pharmacology and Experimental Therapeutics*, **196**, 97–106.

Dum, J., Bläsig, J., Meyer, G. & Herz, A. (1979). Opiate antagonist–receptor interaction unchanged by acute or chronic opiate treatment. *European Journal of Pharmacology*, **55**, 375–83.

Ehrenpreis, S., Light, I. & Schonbuch, G. H. (1972). Use of the electrically stimulated guinea pig ileum to study potent analgesics. In *Drug Addiction: Experimental Aspects*, ed. J. M. Singh, L. H. Miller & H. Zal, pp. 319–42. New York: Futura.

Fennessy, M. R., Heimans, R. L. H. & Rand, M. J. (1969). Comparison of effect of morphine-like analgesics on transmurally stimulated guinea-pig ileum. *British Journal of Pharmacology*, **37**, 436–49.

Gabella, G. (1981). Structure of muscles and nerves in the gastrointestinal tract. In *Physiology of the Gastrointestinal Tract*, ed. L. R. Johnson, pp. 197–241. New York: Raven Press.

Greenberg, R., Kosterlitz, H. W. & Waterfield, A. A. (1970). The effects of hexamethonium, morphine and adrenaline on the output of acetylcholine from the myenteric plexus – longitudinal muscle preparation of the ileum. *British Journal of Pharmacology*, **40**, 553P–554P.

Griffith, S. G., Meghji, P., Moody, C. J. & Burnstock, G. (1981). 8-Phenyltheophylline: a potent P_1-purinoceptor antagonist. *European Journal of Pharmacology*, **75**, 61–4.

Hammond, M. D., Schneider, C. & Collier, H. O. J. (1976). Induction of opiate tolerance in isolated guinea pig ileum and its modification by drugs. In *Opiates and Endogenous Opioid Peptides*, ed. H. W. Kosterlitz, pp. 169–76. Amsterdam: Elsevier/North-Holland Biomedical Press.

Henderson, G., Hughes, J. & Kosterlitz, H. W. (1972). A new example of a morphine-sensitive neuro-effector junction: adrenergic transmission in the mouse vas deferens. *British Journal of Pharmacology*, **46**, 764–6.

Hitzemann, R. J., Hitzemann, B. A. & Loh, H. H. (1974). Binding of ^{3}H-naloxone in the mouse brain: effect of ions and tolerance development. *Life Sciences*, **14**, 2393–404.

Höllt, V. & Wüster, M. (1978). The opiate receptors. In *Developments in Opiate Research*, ed. A. Herz, pp. 1–65. New York: Marcel Dekker.

Hughes, J., Kosterlitz, H. W. & Leslie, F. M. (1975). Effect of morphine on adrenergic transmission in the mouse vas deferens. Assessment of agonist and antagonist potencies of narcotic analgesics. *British Journal of Pharmacology*, **53**, 371–81.

Huidobro-Toro, J. P. & Way, E. L. (1981). Contractile effect of morphine and related opioid alkaloids, β-endorphin and methionine enkephalin on the isolated colon from Long Evans rats. *British Journal of Pharmacology*, **74**, 681–94.

Johnson, S. M., Westfall, D. P., Howard, S. A. & Fleming, W. W. (1978). Sensitivities of the isolated ileal longitudinal smooth muscle–myenteric plexus and hypogastric nerve–vas deferens of the guinea pig after chronic morphine pellet implantation. *Journal of Pharmacology and Experimental Therapeutics*, **204**, 54–66.

Karras, P. J. & North, R. A. (1981). Acute and chronic effects of opiates on single neurones of the myenteric plexus. *Journal of Pharmacology and Experimental Therapeutics*, **217**, 70–80.

Klee, W. A. & Nirenberg, M. (1974). A neuroblastoma × glioma hybrid cell line with morphine receptors. *Proceedings of the National Academy of Sciences, USA*, **71**, 3474–7.

Klee, W. A. & Streaty, R. A. (1974). Narcotic receptor sites in morphine-dependent rats. *Nature, London*, **248**, 61–3.

Kosterlitz, H. W. & Watt, A. J. (1968). Kinetic parameters of narcotic agonists and antagonists, with particular reference to N-allylnoroxymorphone (naloxone). *British Journal of Pharmacology and Chemotherapy*, **33**, 266–76.

Lampert, A., Nirenberg, M. & Klee, W. A. (1976). Tolerance and dependence evoked by an endogenous opiate peptide. *Proceedings of the National Academy of Sciences, USA*, **73**, 3165–7.

Lee, C. Y., Stolman, S., Akera, T. & Brody, T. M. (1973). Saturable binding of ^{3}H-dihydromorphine to rat brain tissue *in vitro*. Characteristics and effect of morphine pretreatment. *Pharmacologist*, **15**, 202.

Llorens, C., Martres, M. P., Baudry, M. & Schwartz, J. C. (1978). Hypersensitivity to noradrenaline in cortex after chronic morphine: relevance to tolerance and dependence. *Nature, London*, **274**, 603–5.

Lord, J. A. H., Waterfield, A. A., Hughes, J. & Kosterlitz, H. W. (1977). Endogenous opioid peptides: multiple agonists and receptors. *Nature, London*, **267**, 495–9.

Margules, D. L. (1979). Beta-endorphin and endoloxone: hormones of the autonomic nervous system for the conservation of expenditure of bodily resources and energy in anticipation of famine or feast. *Neuroscience and Biobehavioural Reviews*, **33**, 155–62.

Morita, K. & North, R. A. (1981). Clonidine activates membrane potassium conductance in myenteric neurones. *British Journal of Pharmacology*, **74**, 419–28.

Nicoll, R. A., Alger, B. E. & Jahr, C. E. (1980). Enkephalin blocks inhibitory pathways in the vertebrate CNS. *Nature, London*, **287**, 22–5.

Nicoll, R. A., Siggins, G. R., Ling, N., Bloom, F. E. & Guillemin, R. (1977). Neuronal actions of endorphins and enkephalins among brain regions: a comparative microiontophoretic study. *Proceedings of the National Academy of Sciences, USA*, **74**, 2584–8.

North, R. A. & Karras, P. J. (1978). Opiate tolerance and dependence induced *in vitro* in single myenteric neurones. *Nature, London*, **272**, 73–5.

North, R. A. & Tonini, M. (1977). The mechanism of action of narcotic analgesics in the guinea-pig ileum. *British Journal of Pharmacology*, **61**, 541–9.

North, R. A. & Vitek, L. V. (1980). The effect of chronic morphine treatment on excitatory junction potentials in the mouse vas deferens. *British Journal of Pharmacology*, **68**, 399–405.

Paton, W. D. M. (1957). The action of morphine and related substances on contraction and on acetylcholine output of coaxially stimulated guinea-pig ileum. *British Journal of Pharmacology and Chemotherapy*, **12**, 119–27.

Paton, W. D. M. (1969). A pharmacological approach to drug dependence and drug tolerance. In *Scientific Basis of Drug Dependence (a Biological Council Symposium)*, ed. H. Steinberg, pp. 31–47. London: J. & A. Churchill.

Paton, W. D. M. & Zar, M. A. (1968). The origin of acetylcholine released from guinea-pig intestine and longitudinal muscle strips. *Journal of Physiology*, **194**, 13–33.

Pert, C. B. & Snyder, S. H. (1976). Opiate receptor binding – enhancement by opiate administration *in vivo*. *Biochemical Pharmacology*, **25**, 847–53.

Rae, G. A., Neto, J. P. & De Moraes, S. (1977). Noradrenaline supersensitivity of the mouse vas deferens after long-term treatment with morphine. *Journal of Pharmacy and Pharmacology*, **29**, 310–12.

Schaumann, W. (1957). Inhibition by morphine of the release of acetylcholine from the intestine of the guinea-pig. *British Journal of Pharmacology and Chemotherapy*, **12**, 115–18.

Schulz, R., Faase, E., Illes, P. & Wüster, M. (1980*a*). Development of opiate tolerance/dependence in the guinea-pig myenteric plexus and the mouse vas deferens. In *Endogenous and Exogenous Opiate Agonists and Antagonists*, ed. E. L. Way, pp. 135–8. Oxford: Pergamon Press.

Schulz, R. & Herz, A. (1976). Aspects of opiate dependence in the myenteric plexus of the guinea-pig. *Life Sciences*, **19**, 1117–28.

Schulz, R., Wüster, M. & Herz, A. (1979). Supersensitivity to opioids following the chronic blockade of endorphin action by naloxone. *Naunyn-Schmiedeberg's Archives of Pharmacology*, **306**, 93–6.

Schulz, R., Wüster, M., Krenss, H. & Herz, A. (1980*b*). Selective development of tolerance without dependence in multiple opiate receptors of mouse vas deferens. *Nature, London*, **285**, 242–3.

Sharma, S. K., Klee, W. A. & Nirenberg, M. (1975*a*). Dual regulation of adenylate cyclase accounts for narcotic dependence and tolerance. *Proceedings of the National Academy of Sciences, USA*, **72**, 3092–6.

Sharma, S. K., Klee, W. A. & Nirenberg, M. (1977). Opiate-dependent modulation of adenylate cyclase. *Proceedings of the National Academy of Sciences, USA*, **74**, 3365–9.

Sharma, S. K., Nirenberg, M. & Klee, W. A. (1975*b*). Morphine receptors as regulators of adenylate cyclase activity. *Proccedings of the National Academy of Sciences, USA*, **72**, 590–4.

Syapin, P. J. & Rickman, D. W. (1981). Benzodiazepine receptor increase following repeated pentylenetetrazole injections. *European Journal of Pharmacology*, **72**, 117–20.

Tokimasa, T., Morita, K. & North, A. (1981). Opiates and clonidine prolong calcium-dependent after-hyperpolarizations. *Nature, London*, **294**, 162–3.

Traber, J., Fischer, K., Latzin, S. & Hamprecht, B. (1975). Morphine antagonises action of prostaglandin in neuroblastoma and neuroblastoma × glioma hybrid cells. *Nature, London*, **253**, 120–2.

Way, E. L., Loh, H. H. & Shen, F. H. (1969). Simultaneous quantitative assessment of morphine tolerance and physical dependence. *Journal of Pharmacology and Experimental Therapeutics*, **167**, 1–8.

Wei, E. & Loh, H. (1976). Physical dependence on opiate-like peptides. *Sciences*, **193**, 1262–3.

Zieglgänsberger, W., French, E. D., Siggins, G. R. & Bloom, F. E. (1979). Opioid peptides may excite hippocampal pyramidal neurons by inhibiting adjacent inhibitory interneurons. *Science*, **205**, 415–17.

CLIFFORD R. ELCOMBE

The induction of hepatic cytochrome(s) P_{450}: an adaptive response?

A major role for the liver is to metabolise exogenous chemicals (xenobiotics). These xenobiotics may be derived from the natural diet (e.g. alkaloids, flavones) or may be contaminants, present either intentionally (drugs, food additives) or unintentionally (pesticides, pollutants, etc.). The liver is capable of metabolising such chemicals by a wide range of chemical mechanisms. Classically, these biotransformation pathways have been termed phase I (pre-synthetic) and phase II (synthetic) reactions (Williams, 1959; Parke, 1968). Phase I involves oxidation and hydrolytic reactions, while phase II involves the conjugation of metabolites with endogenous compounds such as glucuronic acid and sulphate (Table 1). In general, xenobiotics are relatively lipid-soluble and the function of the hepatic xenobiotic metabolising enzymes is to increase the water solubility of these exogenous compounds, enabling rapid excretion from the body.

Of primary importance are the oxidative enzyme systems located in the hepatic endoplasmic reticulum, known collectively as microsomal monooxygenases or mixed function oxidases. These are described in detail below.

The administration of a variety of chemicals to animals results in extensive liver enlargement, frequently associated with a marked proliferation of smooth endoplasmic reticulum and concomitant increases in xenobiotic metabolising enzymes. Such changes have frequently been termed 'adaptive' (Schulte-Hermann, 1979). The *Longman English Dictionary* defines adaptive as 'tending or able to adapt' and defines adapt as 'to put in harmony with changed circumstances, or, 'to make more suitable by altering'. This definition implies a beneficial aspect to adaptation.

This article describes the process of induction of microsomal monooxygenase activity and its toxicological implications. Furthermore the author attempts to assess whether such enzyme induction is in general beneficial and hence correctly termed 'adaptive'.

Multiple forms and induction of cytochromes P_{450}

The presence of a complex multi-enzyme monooxygenase system within the hepatic endoplasmic reticulum has been demonstrated in several mammalian species (Estabrook, Gillette & Liebman, 1972; Ullrich *et al.*, 1977; Coon *et al.*, 1980*a, b*). These monooxygenase systems exhibit an unusually low substrate specificity and are responsible for the oxidative metabolism of many exogenous and endogenous chemicals (e.g. drugs, pesticides, food additives, steroids, fatty acids). The oxygen-activating enzyme(s) involved in the microsomal monooxygenase reactions is a haem-containing protein and has been named cytochrome P_{450} because the carboxyferrocytochrome has an optical absorbance maximum at 450 nm. Hepatic microsomal monooxygenase activities may be increased by a wide range of drugs and chemicals. Two types of stimulation of monooxygenation were initially observed, and shown to be due to the *de novo* synthesis of cytochrome P_{450}. The first type of induction is elicited by a wide variety of chemicals such as the barbiturates, phenylbutazone and 2,2-bis(*p*-chlorphenyl)-1,1,1-trichloroethane (DDT), and is characterised by the stimulation of metabolism of a broad range of substrates (e.g. the *N*-demethylation of benzphetamine and ethylmorphine). A second class of inducing agents is typified by the polycyclic aromatic hydrocarbons (e.g. 3-methylcholanthrene, benzo(a)pyrene, β-naphthoflavone) and 2,3,7,8-tetrachloro-*p*-dibenzodioxin (TCDD). This class of inducer stimulates a narrower range of substrate metabolism (e.g. benzo(a)pyrene hydroxylation and ethoxyresorufin-*O*-de-ethylation). Furthermore, these agents increase the synthesis of a cytochrome with an absorbance maximum at 448 nm (cytochrome P_{448} or P_{450}^1). Because of the difference in spectral characteristics the term cytochrome(s) P_{450} will be used generically.

Table 1. *Major pathways of xenobiotic metabolism*

Phase I reactions		
1	Oxidation	Cytochrome P_{450}
		Amine oxidase
2	Reduction	Azoreductase
		Nitroreductase
3	Hydrolysis	Esterases
4	Hydration	Epoxide hydrolase
Phase II reactions		
1	Conjugation	Glucuronyl transferase
		Glutathione transferase
		Sulphotransferase
		Acetylase

More recently other chemicals have been found to fit into these categories of inducing agents. For example, non-planar polychlorinated biphenyls (substituted in the 2 or 6 positions) induce in a similar, if not identical manner, to barbiturates, while the planar polychlorinated biphenyls induce in a manner similar to the polycyclic aromatic hydrocarbons. Fig. 1 illustrates representative structures of coplanar and non-coplanar polychlorinated biphenyls (PCBs).

Using these compounds as model inducing agents, many quantitative and qualitative distinctions in the induction of monooxygenase activity can be made.

The administration of either pure PCB congeners or commercial mixtures of PCBs (Aroclors 1254 and 1242) to rats results in increased concentrations of microsomal cytochrome(s) P_{450} (Table 2). Furthermore, it should be noted that the administration of the PCB mixtures or the coplanar congener (3,4,3',4'-tetrachlorobiphenyl) resulted in a shift of the absorbance maximum to approximately 448 nm. The pure non-coplanar congener (2,4,5,2',4',5'-hexachlorobiphenyl) had no significant effect upon the absorbance maximum of the carboxyferrocytochrome. Monooxygenase activities were also differentially induced by these agents. Table 3 demonstrates that the PCB mixtures increased the N-demethylation of benzphetamine, and the O-de-ethylations of ethoxycoumarin and ethoxyresorufin. However, the coplanar congener (3,4,3',4'-tetrachlorobiphenyl) only induced ethoxycoumarin- and ethoxyresorufin-O-deethylations, while the non-coplanar congener (2,4,5,2',4',5'-hexachlorobiphenyl) only increased benzphetamine-N-demethylation and ethoxycoumarin-O-deethylation. These differences in the profile of induced substrate metabolism together with alterations in the absorbance maximum of the carboxyferrocytochrome suggest qualitative changes in the subpopulations of cytochrome(s) P_{450}. Sodium dodecylsulphate–polyacrylamide gel

Fig. 1. Representative structures of polychlorinated biphenyls. 3,4,3',4'-Tetrachlorobiphenyl is coplanar (unsubstituted in 'bridge' position), while 2,4,5,2',4',5'-hexachlorobiphenyl is non-coplanar.

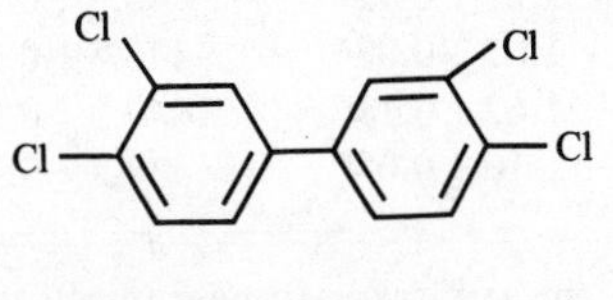

3, 4, 3", 4'-Tetrachlorobiphenyl

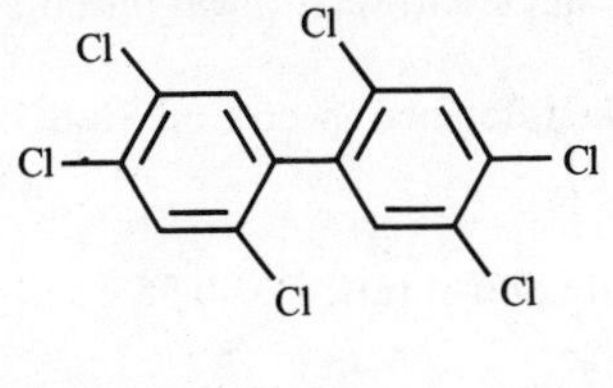

2, 4, 5, 2', 4', 5'-Hexachlorobiphenyl

electrophoresis (SDS-PAGE) of hepatic microsomes obtained from rats treated with various inducing agents show marked differences in the protein profiles in the 45000 to 60000 dalton region (Fig. 2). The protein profile induced by 3,4,3′,4′-tetrachlorobiphenyl is similar to that induced by 3-methylcholanthrene, while the profile induced by 2,4,5,2′,4′,5′-hexachlorobiphenyl is similar to that induced by phenobarbital. Such changes are indeed due to alterations in cytochrome(s) P_{450} as shown by purification and reconstitution studies (Lu & West, 1980).

The early division of inducing agents into two classes is now recognised to be an oversimplification. At least three other distinct types of inducing agents are now known; these are typified by isosafrole, pregnenolone-16α-carbonitrile and ethanol.

Hence it is apparent that cytochrome(s) P_{450} is not a single highly non-specific enzyme, but a heterogenous mixture of several enzymes each having its own range of substrate specificities.

Strain and species differences in the induction of cytochrome(s) P_{450}

Marked strain and species differences exist in the response of animals to inducing agents. Table 4 shows that, even in mammals, interspecies differences in induction of monooxygenase activity by 3-methylcholanthrene (3-MC) are apparent. After administration of 3-MC to mouse and rat marked increases in arylhydrocarbon (benzo(a)pyrene) hydroxylation, (AHH), 2-acetylaminofluorene-N-hydroxylation (AAF-N-OHase) and biphenyl-2-

Table 2. *Cytochrome P_{450} content of hepatic microsomes from variously induced female rats*

Treatment	n	Total P_{450} (nmol mg^{-1} protein)[a]	λ_{max} (nm)
Corn oil	15	0.67 ± 0.02[b]	449.8 ± 0.04[b]
Aroclor 1254	4	2.27 ± 0.12[c]	448.3 ± 0.06[c]
Aroclor 1242	4	1.05 ± 0.08[c]	448.5 ± 0.07[c]
2,4,5,2′,4′,5′-Hexachlorobiphenyl	4	1.62 ± 0.11[c]	449.9 ± 0.05
3,4,3′,4′-Tetrachlorobiphenyl	4	1.10 ± 0.08[c]	448.2 ± 0.10[c]

Female rats (180–200 g) were treated with a single intraperitoneal injection of each agent (150 mg kg^{-1}) and killed 4 days later.

[a] Total P_{450} determined by carbon monoxide difference spectra of Na_2SO_4-reduced microsomal suspensions ($\epsilon = 91$ mM^{-1} cm^{-1}).
[b] Values are mean $\pm$ S.E.
[c] Significantly different from corn-oil-treated control rats, $P < 0.05$.

Table 3. Induction of hepatic microsomal monooxygenase activity in female rats by various polychlorinated biphenyl congeners

Treatment	n	BeND[a] (nmol min^{-1} mg^{-1})	ECOD[b] (nmol min^{-1} mg^{-1})	EROD[c] (nmol min^{-1} mg^{-1})
Corn oil	16	2.50 ± 0.29[d]	1.37 ± 0.18[d]	0.68 ± 0.04[d]
Aroclor 1254	4	17.1 ± 0.2[e]	15.3 ± 1.9[e]	21.5 ± 4.8[e]
Aroclor 1242	4	6.42 ± 1.67[e]	10.4 ± 1.9[e]	14.5 ± 1.5[e]
2,4,5,2′,4′,5′-Hexachlorobiphenyl	4	20.6 ± 1.0[e]	4.60 ± 0.10[e]	0.98 ± 0.20
3,4,3′,4′-Tetrachlorobiphenyl	4	1.48 ± 0.18	10.1 ± 0.6[e]	15.7 ± 3.0[e]

Female rats (180–200 g) were treated with a single intraperitoneal injection of each agent (150 mg kg^{-1}) and killed 4 days later.
[a] Benzphetamine-N-demethylation.
[b] Ethoxycoumarin-O-deethylation.
[c] Ethoxyresorufin-O-deethylation.
[d] Values are mean ± S.E.
[e] Significantly different from corn-oil-treated control rats, $P < 0.05$.

hydroxylation (BIP-2-OHase) are observed. In the hamster little effect upon AHH is seen, while in the guinea-pig 3-MC had no significant effect upon any of the monooxygenase activities measured.

Even larger contrasts are noticeable when mammalian/non-mammalian comparisons are made. For example fish are refractile to the effects of barbiturate-type inducing agents. This is illustrated by again using the PCBs as model inducing agents (Table 5). The non-coplanar PCB congeners

Fig. 2. Sodium dodecyl sulphate–polyacrylamide gel electrophoresis of female rat hepatic microsomes. Forty-five micrograms of microsomal protein was applied to each well. The microsomes were obtained from rats administered: A, isosafrole; B, 3,4,3′,4′-tetrachlorobiphenyl; C, 2,4,5,2′,4′5′-hexachlorobiphenyl; D, Aroclor 1254; E, Aroclor 1242; F, 3-methylcholanthrene; G, phenobarbital; H, corn oil (vehicle control). The gel was stained with Coomassie Blue. The 45000 to 60000 dalton region is indicated.

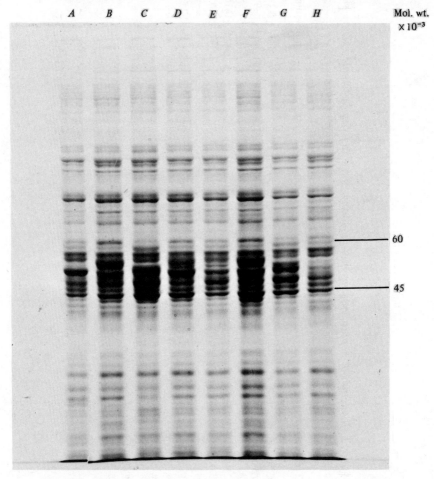

(2,4,2′,4′-tetra and 2,4,5,2′,4′,5′-hexa) had no effect upon benzphetamine-N-demethylation (BeND) or ethoxycoumarin-O-deethylation (ECOD), while the PCB mixtures and the coplanar PCB congeners (3,4,5,3′,4′-penta and 3,4,3′,4′-tetra) greatly increased ECOD and EROD. Furthermore only the mixed PCBs and the coplanar congeners altered the microsomal protein profiles (Fig. 3), inducing a novel protein of molecular weight 57000. (For a review see Lech, Vodicnik & Elcombe, 1982).

Strain differences in the susceptibility of mice to induction of cytochrome P_{448} by polycyclic aromatic hydrocarbons are well documented. Fig. 4 shows the difference in stimulation of AHH (a P_{448}-dependent activity) in various strains of mice. The basal (uninduced) AHH activities of the C57BL/6N and the DBA/2N mice were very similar; however, after treatment of the animals with 3-MC a large increase in AHH was seen in the B6 mouse while no effect was observed in the D2 mouse. All individuals of the F_1 hybrid of these strains were responsive to 3-MC induction, but when the F_1 hybrid was back-crossed with the non-responsive D2 strain, a 50:50 split in responsiveness was observed. This strain difference has been very well characterised and appears to be due to a defective cytosolic receptor for 3-MC. An extensive description of the genetics of induction has been presented by Nebert et al. (1981).

Effect of induction on the metabolism and toxicity of chemicals

Inducers may increase the rate of their own in vivo metabolism as well as the rate of metabolism of other agents. Fig. 5 illustrates that multiple dosing of hexobarbital to dogs increases the rate of elimination of hexobarbital from the blood. Furthermore, after cessation of multiple dosing the rate of hexobarbital elimination returns to normal. Similarly phenobarbital pre-

Table 4. *3-Methylcholanthrene induction of monooxygenase activities in liver microsomes of mice, rats, hamsters and guinea-pigs*

Species	Enzyme activity (% of respective control)		
	AHH[a]	AAF-N-OHASE[b]	BIP-2-OHASE[c]
Mouse	550	570	430
Rat	990	300	> 580
Hamster	120	450	210
Guinea-pig	140	135	110

Data adapted from Thorgeirsson et al. (1979).
[a] Arylhydrocarbon hydroxylation.
[b] 2-Acetylaminofluorene-N-hydroxylation.
[c] Biphenyl-2-hydroxylation.

Table 5. Induction of hepatic microsomal monooxygenase activity in the rainbow trout by various polychlorinated biphenyl congeners

Treatment	BeND[a] (nmol min^{-1} mg^{-1})	ECOD[b] (nmol min^{-1} mg^{-1})	EROD[c] (nmol min^{-1} mg^{-1})
Corn oil	1.05[d]	0.023 ± 0.003[e]	0.025 ± 0.009[e]
Aroclor 1254	0.51	—	0.365 ± 0.100[f]
Aroclor 1242	1.40	0.186 ± 0.053[f]	0.608 ± 0.379[f]
2,4,2',4'-Tetrachlorobiphenyl	1.08	0.045 ± 0.008	0.085 ± 0.059
3,4,3',4'-Tetrachlorobiphenyl	1.37	0.148 ± 0.015[f]	0.519 ± 0.101[f]
3,4,5,3',4'-Pentachlorobiphenyl	0.47	0.647 ± 0.078[f]	0.356 ± 0.051[f]
2,4,5,2',4',5'-Hexachlorobiphenyl	1.20	0.030 ± 0.001	0.010 ± 0.007

Rainbow trout (about 100 g) were treated with a single intraperitoneal injection of each agent (150 mg kg^{-1}) and killed 5 days later.
[a] Benzphetamine-N-demethylation.
[b] Ethoxycoumarin-O-deethylation.
[c] Ethoxyresorufin-O-deethylation.
[d] Values obtained using pooled microsomes from four fish.
[e] Values are mean ± s.e. ($n = 8$).
[f] Significantly different from corn-oil-treated control fish, $P < 0.05$.

treatment of dogs also increases the rate of elimination of hexobarbital. Inducers may also increase the rate of plasma clearance of structurally unrelated compounds. For example, treatment of rats with phenobarbital, benzo(a)pyrene or Aroclor 1254 dramatically increases the rate of elimination of caffeine from the blood (Fig. 6).

Inducers not only quantitatively alter metabolism, but may also lead to different metabolite profiles. Cytochrome P_{448} inducers (e.g. 3-MC) increase the 7,8- and 9,10-epoxidation of benzo(a)pyrene and the 2-hydroxylation of biphenyl, while cytochrome P_{450} inducers (e.g. phenobarbital) preferentially increase the 3-hydroxylation and the 4,5-epoxidation of benzo(a)pyrene and the 4-hydroxylation of biphenyl (Fig. 7a and b).

Alterations in metabolic pathways may be responsible for changes in the toxicity of chemicals. For example, if toxicity is mediated through a metabolite rather than the parent compound, induction of a cytochrome

Fig. 3. Sodium dodecyl sulphate–polyacrylamide gel electrophoresis of rainbow trout hepatic microsomes. Forty-five micrograms (A–F) or 90 μg (G–L) of microsomal protein was applied to each well. The microsomes were obtained from fish pretreated with: A and G, corn oil (vehicle control); B and H, 2,4,2′,4′-tetrachlorobiphenyl; C and I, 2,4,5,2′,4′5′-hexachlorobiphenyl; D and J, 3,4,3′,4′-tetrachlorobiphenyl; E and K, Aroclor 1242; F and L, Aroclor 1254. The gel was stained with Coomassie Blue and the 57 000 dalton protein is indicated.

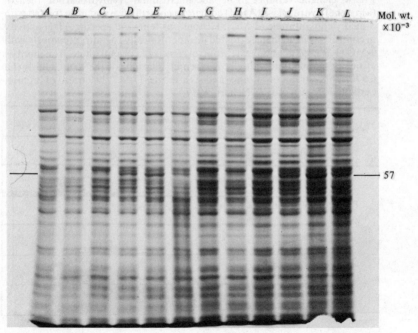

which preferentially forms that specific metabolite will increase toxicity. Alternatively, induction of a cytochrome which metabolises the parent compound to a different metabolite may decrease the toxicity.

The strain differences in the responsiveness of mice to 3-MC induction have been utilised to study such induction-mediated modifications of toxicity.

The administration of paracetamol (500 mg kg^{-1} body wt) to C57BL/6N mice results in mild to moderate hepatic necrosis and no mortality (Table 6). No hepatic effects are seen at a lower dose of 200 mg kg^{-1} body wt. However, prior treatment of the mice with 3-MC, and subsequent administration of paracetamol (500 mg kg^{-1} body wt), results in 100% mortality. Lower doses of paracetamol lead to more severe hepatic necrosis than seen in the uninduced animals. In contrast, the toxicity of paracetamol in the non-responsive DBA/2N mouse is barely affected by 3-MC administration.

Conversely, the central nervous system toxicity of lindane is decreased by induction of cytochrome P_{448}. Table 7 illustrates that both C57BL/6N and DBA/2N mice die within 12 hours after the administration of lindane. However, induction of cytochrome P_{448} by 3-MC decreases the toxicity of lindane in the C57BL/6N mouse (27 animals alive after 5 days), whereas such induction has no effect upon survival in the DBA/2N mouse.

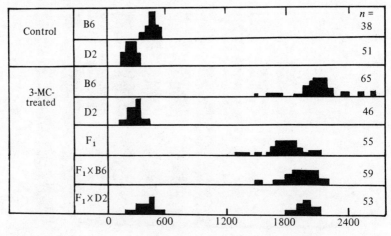

Fig. 4. Genetic variance in hepatic microsomal arylhydrocarbon (benzo-(a)pyrene) hydroxylase (AHH) activity in control mice and mice treated with 3-methylcholanthrene (3-MC). Histograms represent specific AHH activity in control mice and in mice treated intraperitoneally 24 hours previously with 3-MC (100 mg kg^{-1} body wt); controls received corn oil alone. The number of mice examined individually is given at the right of the figure. B6 is the C57BL/6N strain of mouse and D2 is the DBA/2N strain of mouse. (Adapted from Kouri & Nehert, 1977.)

Various monooxygenase inducers have been shown to modify chemical carcinogenesis (Wattenberg, 1978). Aflatoxin B_1 (a metabolite of the mould *Aspergillus flavus*) is a potent carcinogen; the rainbow trout is very susceptible to the development of hepatocellular carcinoma when treated with contaminated diet. When trout diet was supplemented with aflatoxin B_1 (6 parts per billion) the fish developed a 70% hepatic tumour incidence after one year (Table 8). Co-administration of Aroclor 1254 decreased the incidence of aflatoxin B_1-elicited tumours to 30%. It is interesting to speculate that the novel cytochrome P_{450} induced by Aroclor 1254 in the trout (see above) is less effective than the constitutive cytochrome at metabolising aflaxtoxin B_1 to the proximate carcinogen (aflatoxin-2,3-epoxide). Evidence for this hypothesis has been presented by Stott & Sinnhuber (1978) who have shown that PCB-induced trout liver metabolises aflatoxin B_1 to mutagenic (and potentially carcinogenic) metabolites to a lesser extent than control trout liver microsomes.

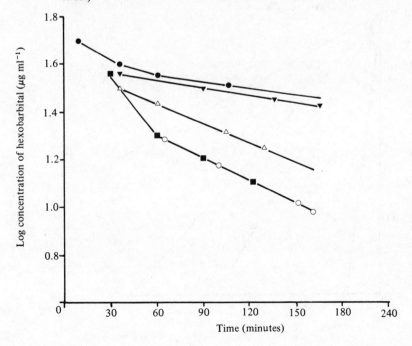

Fig. 5. The elimination of hexobarbital from the blood of dogs. Hexobarbital (30 mg kg^{-1} day^{-1}) was administered by intravenous injection on five successive days. After each injection the rate of fall of hexobarbital concentration in the blood was measured. ●—●, first day; △—△, second day; ■—■, third day; ○—○, fifth day; ▼—▼, 28 days later. (Adapted from Remmer, 1968.)

The effect of induction of cytochrome(s) P_{450} on the metabolism of steroid hormones and possible physiological implications

Induction of monooxygenase activity by compounds such as phenobarbitone, polybrominated biphenyls (PBBs) and PCBs has resulted in the increased rate of metabolism of steroid hormones *in vitro* and *in vivo*. Such stimulation of sex hormone metabolism has been correlated with the modified action of exogenously administered steroids such as progesterone, testosterone and 17β-oestradiol.

The PCBs and PBBs are potent monooxygenase inducers and also have effects upon the reproductive system (Subcommittee on the Health Effects of PCBs and PBBs, 1978). Rodents treated with PCBs or PBBs have shown delayed vaginal opening, lengthened oestrous cycles, uterine atrophy and reduced plasma progesterone concentrations. Similarly rhesus monkeys fed a diet containing PBBs had lengthened menstrual cycles. Women with Yusho disease, caused by consumption of PCB-contaminated rice oil, had menstrual

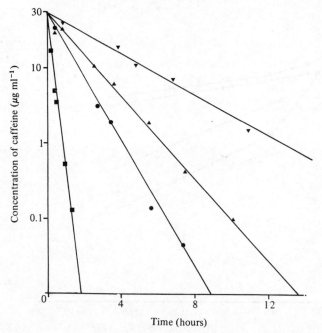

Fig. 6. The disappearance of caffeine from the plasma of rats pretreated with various environmental substances. Rats were dosed orally each morning for three consecutive days with: ▼—▼, corn oil (vehicle control); ▲—▲, phenobarbital; ●—●, benzo(a)pyrene; or ■—■, Aroclor 1254. On day 4 each rat received an intravenous injection of caffeine (20 mg kg^{-1}) and the plasma levels of caffeine were determined at various time intervals. (Adapted from Welch, Hsu & DeAngelis, 1977.)

cycle irregularities, dysmenorrhoea and altered serum ketosteroid concentrations.

It has been suggested that the toxicological effects of the PCBs and PBBs noted above are due to the induction of cytochrome(s) P_{450} and concomitant alterations in steroid metabolism.

Steroids are essential neonatally for the imprinting of many sex-differentiated functions including male sexual behaviour, the pattern of gonadotrophin secretion and patterns of certain drug- and steroid-metabolising enzymes (Skett & Gustafsson, 1979). Hence it is plausible that if neonatal steroid hormone metabolism is altered (either qualitatively or quantitatively), changes in neonatal imprinting may occur. Chemicals such as the PCBs may be transferred from an exposed mother to her litter via the milk. Fig. 8 illustrates

Fig. 7. Alterations in the metabolite profiles of benzo(a)pyrene and biphenyl due to induction. (a) The various chemical structures indicated are: A, benzo(a)pyrene; B, 3-hydroxy-benzo(a)pyrene; C, benzo(a)pyrene-4,5-oxide; D, benzo(a)pyrene-7,8-oxide; E, benzo(a)pyrene-7,8-diol; F, benzo(a)pyrene-7,8-diol-9,10-oxide. (b) The chemical structures shown are: A, biphenyl; B, 2-hydroxybiphenyl; C, 4-hydroxybiphenyl.

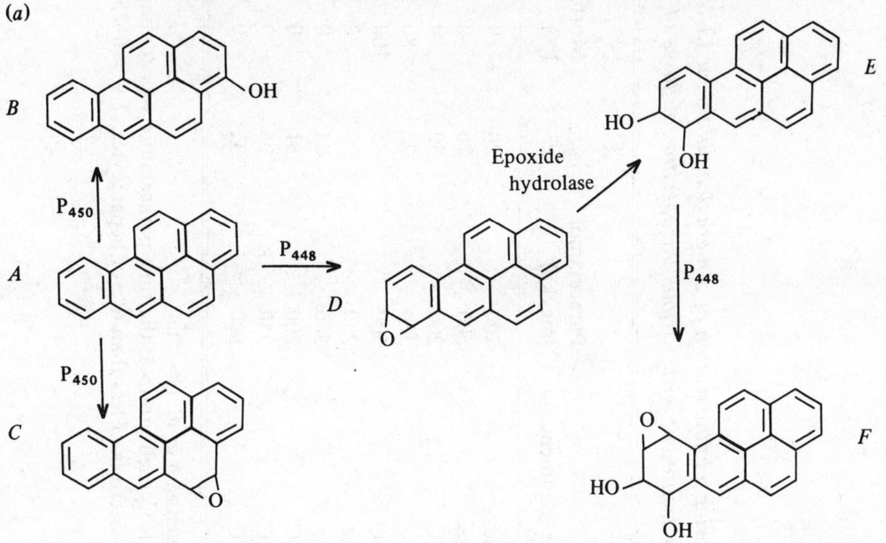

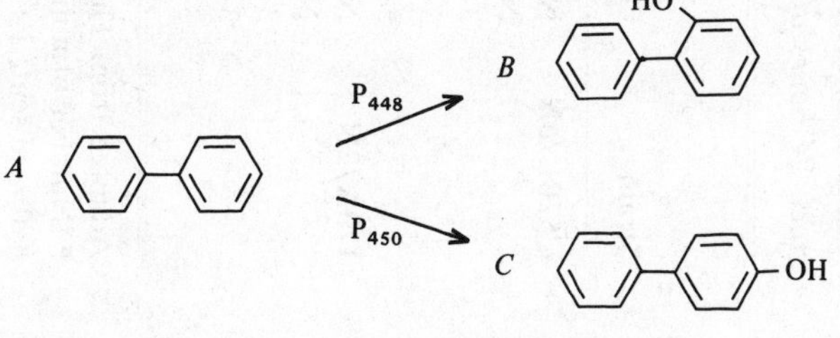

Table 6. *Effect of prior administration of 3-methylcholanthrene (3-MC) on extent of hepatic necrosis in C57BL/6N and DBA/2N mice 24 hours after treatment with various intraperitoneal doses of paracetamol (acetaminophen)*

Strain	Prior treatment	Paracetamol (mg/kg)	n^a	Mortality (%)	Extent of necrosis[b]				
					0	1+	2+	3+	4+
C57BL/6N	None	200	20	0	100	—	—	—	—
	None	500	26	0	—	—	38	52	10
	3-MC	200	30	10	10	—	25	45	20
	3-MC	350	28	80	—	—	—	33	67
	3-MC	500	35	100	—	—	—	—	—
DBA/2N	None	200	20	0	100	—	—	—	—
	None	500	24	0	—	5	70	25	—
	3-MC	200	31	0	100	—	—	—	—
	3-MC	350	25	0	—	—	40	60	—
	3-MC	500	32	0	—	—	48	72	—

Adapted from Thorgeirsson *et al.* (1975).
[a] Number of animals in each group at the time paracetamol was administered.
[b] 0 = absent; 1+ = necrosis of less than 6% of hepatocytes; 2+ = 6–25%; 3+ = 25–50%; 4+ = more than 50% of hepatocytes necrotic.

Table 7. *Prevention of lindane toxicity in mice by previous treatment with 3-methylcholanthrene (3-MC)*

Strain	AHH phenotype	Previous treatment	Number of animals dying between				Animals still alive after 5 days
			0 and 3 hours	4 and 6 hours	7 and 9 hours	10 and 12 hours	
C57BL/6N		Corn oil	2	6	12	9	1
DBA/2N		Corn oil	4	10	8	8	0
C57BL/6N	Responsive	3-MC	0	1	1	1	27
DBA/2N	Non-responsive	3-MC	9	8	8	5	0

Adapted from Robinson *et al.* (1975).
Each mouse received 3-MC (80 mg kg^{-1} intraperitoneally) or corn oil alone 48 hours before intraperitoneal administration of 300 mg kg^{-1} lindane (each group contained 30 mice).

an experimental situation where precocious induction of monooxygenase activity is elicited in pups from mice injected *before* pregnancy with 2,4,5,2',4',5'-hexachlorobiphenyl. Studies have shown that this precocious induction is related to the transfer of PCB during suckling (Vodicnik & Lech, 1980; Vodicnik, Elcombe & Lech, 1980).

Table 8. *Hepatocellular carcinoma incidence in rainbow trout fed four diets*

Diet	Tumour incidence[a]
Control	0/68 (0%)
Aflatoxin B_1 (6 p.p.b.)	26/37 (70%)
Aroclor 1254 (100 p.p.m.)	0/39 (0%)
Aflatoxin B_1 + Aroclor 1254 (6 p.p.b. + 100 p.p.m.)	14/46 (30%)

Modified from Hendricks *et al.* (1977).
[a] Hepatocellular carcinoma incidence after 12 months on diet.

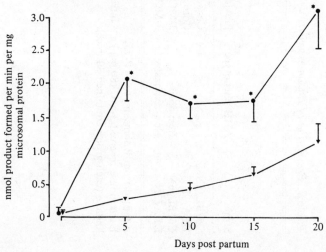

Fig. 8. Precocious induction of hepatic microsomal ethoxycoumarin-*O*-deethylation in the offspring of mice treated with 2,4,5,2',4',5'-hexachlorobiphenyl. Female mice were administered a single intraperitoneal injection of 2,4,5,2',4',5'-hexachlorobiphenyl (100 mg/kg body wt) 14 days before mating. Hepatic microsomal ethoxycoumarin-*O*-deethylation was determined in the offspring of these animals from birth to 20 days *post partum*. ●—●, offspring from 2,4,5,2',4',5'-hexachlorobiphenyl-treated mice; ▼—▼, offspring from corn-oil-treated (control) mice. Values are mean±S.E.; asterisks indicate values significantly different from control. (Adapted from Vodicnik *et al.*, 1980.)

Summary and conclusions

The liver can alter in response to an enhanced functional load by an increase in specific enzymes, multiplication of organelles and by growth of the whole organ. The definition of these responses as adaptive is usually not based on precise information on the underlying molecular and cellular events and is hence often somewhat hypothetical. Furthermore, it should be remembered that these 'adaptive changes' are not necessarily beneficial for the organism. In this context it should be noted that the changes described above are reversible, and the time course of reversibility is correlated with the rate of elimination of the inducing agent.

Induction of microsomal monooxygenases has been shown to lead to both beneficial and detrimental effects for the animal, the type of effect depending on the toxin and the inducer. Hence great care should be taken in extrapolating data from one inducer to another.

Generally, induction of microsomal monooxygenase activity may be regarded as a defence mechanism for increasing the metabolism and elimination of the inducing agent. However, secondary effects may lead to hazardous results at particular times of the animal's life – for example when exposed to several chemicals simultaneously or during development. Hence the primary event of induction would appear to be correctly termed adaptive, the detrimental effects of induction being due to secondary events.

References

Coon, M. J., Conney, A. H., Estabrook, R. W., Gelboin, H. V., Gillette, J. R. & O'Brien, P. J. (eds.) (1980a). *Microsomes, Drug Oxidations and Chemical Carcinogenesis*, vol. 1. New York & London: Academic Press.

Coon, M. J., Conney, A. H., Estabrook, R. W., Gelboin, H. V., Gillette, J. R. & O'Brien, P. J. (eds.) (1980b). *Microsomes, Drug Oxidations and Chemical Carcinogenesis*, vol. 2. New York & London: Academic Press.

Estabrook, R. W., Gillette, J. R. & Liebman, K. C. (eds.) (1972). *Microsomes and Drug Oxidations*. Baltimore: Williams & Wilkins.

Hendricks, J. D., Putnam, T. P., Bills, D. D. & Sinnhuber, R. O. (1977). Inhibitory effect of a polychlorinated biphenyl (Aroclor 1254) on aflatoxin B_1 carcinogenesis in rainbow trout (*Salmo gairdneri*). *Journal of the National Cancer Institute*, **59**, 1545–51.

Kouri, R. E. & Nebert, R. W. (1977). Genetic regulation of susceptibility to polycyclic-hydrocarbon-induced tumors in the mouse. In *Origins of Human Cancer*, book B, ed. H. H. Hiatt, J. D. Watson & J. A. Winsten, pp. 811–35. New York: Cold Spring Harbor Laboratory.

Lech, J. J., Vodicnik, M. J. & Elcombe, C. R. (1982). Induction of monooxygenase activity in fish. In *Aquatic Toxicology*, vol. 1, ed. L. Weber, pp. 107–48. New York: Raven Press.

Lu, A. Y. H. & West, S. B. (1980). Multiplicity of mammalian microsomal cytochromes P-450. *Pharmacological Reviews*, **31**, 277–95.

Nebert, D. W., Eisen, H. J., Negishi, M., Lang, M. A., Hjelmeland, L. M.

& Okey, A. B. (1981). Genetic mechanisms controlling the induction of polysubstrate monooxygenase (P-450) activities. *Annual Review of Pharmacology and Toxicology*, **21**, 431–62.

Parke, D. (1968). *The Biochemistry of Foreign Compounds*. Oxford: Pergamon Press.

Remmer, H. (1968). Induction of drug-metabolizing enzymes in the endoplasmic reticulums of the liver by treatment with drugs. *German Medical Monthly*, **13**, 53–9.

Robinson, J. R., Felton, J. S., Levitt, R. C., Thorgeirsson, S. S. & Nebert, D. W. (1975). Relationship between 'aromatic hydrocarbon responsiveness' and the survival time in mice treated with various drugs and environmental compounds. *Molecular Pharmacology*, **11**, 850–65.

Schulte-Hermann, R. (1979). Reactions of the liver to injury: adaptation. In *Toxic Injury of the Liver*, part A, ed. E. Farber & M. M. Fisher, pp. 385–444. New York: Marcel Dekker.

Skett, P. & Gustafsson, J. A. (1979). Imprinting of enzyme systems of xenobiotic and steroid metabolism. In *Reviews in Biochemical Toxicology*, vol. 1, ed. E. Hodgson, J. R. Bend & R. M. Philpot, pp. 27–52. New York: Elsevier/North-Holland.

Stott, W. T. & Sinnhuber, R. O. (1978). Trout hepatic enzyme activation of aflatoxin B_1 in a mutagen assay system and the inhibitory effect of PCBs. *Bulletin of Environmental Contamination and Toxicology*, **19**, 35–41.

Subcommittee on the Health Effects of Polychlorinated Biphenyls and Polybrominated Biphenyls (1978). Final Report of the Subcommittee on the Health Effects of Polychlorinated Biphenyls and Polybrominated Biphenyls of the DHEW Committee to Coordinate Toxicology and Related Programs. *Environmental Health Perspectives*, **24**, 129–239.

Thorgeirsson, S. S., Atlas, S. A., Boobis, A. R. & Felton, J. S. (1979). Species differences in the substrate specificity of hepatic cytochrome P-448 from polycyclic hydrocarbon-treated animals. *Biochemical Pharmacology*, **28**, 217–26.

Thorgeirsson, S. S., Felton, J. S. & Nebert, D. W. (1975). Genetic differences in the aromatic hydrocarbon-inducible *N*-hydroxylation of 2-acetylaminofluorene and acetaminophen-produced hepatotoxicity in mice. *Molecular Pharmacology*, **11**, 159–65.

Ullrich, V., Roots, I., Hildebrandt, A., Estabrook, R. W. & Conney, A. H. (eds.) (1977). *Microsomes and Drug Oxidations*. Oxford: Pergamon Press.

Vodicnik, M. J., Elcombe, C. R. & Lech, J. J. (1980). The transfer of 2,4,5,2′,4′,5′-hexachlorobiphenyl to fetuses and nursing offspring. II. Induction of hepatic microsomal monooxygenase activity in pregnant and lactating mice and their young. *Toxicology and Applied Pharmacology*, **54**, 301–10.

Vodicnik, M. J. & Lech, J. J. (1980). The transfer of 2,4,5,2′,4′,5′-hexachlorobiphenyl to fetuses and nursing offspring. I. Disposition in pregnant and lactating mice and their young. *Toxicology and Applied Pharmacology*, **54**, 293–300.

Wattenberg, L. W. (1978). Inhibitors of chemical carcinogenesis. *Advances in Cancer Research*, **21**, 197–226.

Welch, R. M., Hsu, S. Y. & DeAngelis, R. L. (1977). Effect of Aroclor 1254, phenobarbital and polycyclic aromatic hydrocarbons on the plasma clearance of caffeine in the rat. *Clinical Pharmacology and Therapeutics*, **22**, 791–8.

Williams, R. T. (1959). *Detoxication Mechanisms*, 2nd edn. London: Chapman & Hall.

J. C. ELLORY and J. S. GIBSON

Cellular aspects of salinity adaptation in teleosts

Teleostean fish, whether living in seawater at an osmolarity of around 1070 mosmol l^{-1}, or freshwater of 20 to < 1 mosmol l^{-1}, normally maintain their blood osmolarity in the range 270–470 mosmol l^{-1} (see Evans, 1980a, for some values). Further, euryhaline fish, which have the ability to survive transfer from freshwater to seawater (or vice versa) do not alter their blood osmolarity by more than a small fraction (around 50 mosmol l^{-1}) of the osmolarity of seawater. This homeostasis is impressive because fish are not impermeable to salt and water, and consequently have to cope with the influx of water and loss of ions in freshwater, whilst needing to conserve water in a saline environment.

Such a response, where internal ionic concentrations are maintained in the face of marked external variations represents *osmoregulation* and reflects the selective functioning of specialised cell types. The contrasting response, seen for example in invertebrates and in euryhaline amphibians, is to vary internal osmolarity so that there is always a gradient favouring a net influx of water. In this kind of adaptation, known as *osmoconformity*, all cell types will be exposed to the changing plasma tonicity (Prosser, 1973). Thus there will be clear differences in the cellular response to these differing strategies of adaptation.

The physiology of teleostean osmoregulation

From the classical studies of Smith (1930), Keys (1933) and Krogh (1939) there has emerged the generally accepted picture of teleost osmoregulation in freshwater or seawater. In seawater, there will be a passive loss of water and gain of salt across the gills and skin. The major osmoregulatory organs which compensate for this are the gills, which show active salt excretion, and the intestine, where there is active salt absorption followed by water. Subsequently the sodium chloride load is excreted via the gill, resulting in net water uptake. Additionally, the kidney in marine teleosts produces a scanty urine whose osmolarity is close to that of plasma (but fish kidneys cannot produce hypertonic urine). However, the kidney is not a major route for sodium chloride regulation in seawater, renal excretion of divalent cations being more important.

In freshwater there will be a net gradient for water entry and sodium chloride loss. Under these conditions the gills function to absorb sodium chloride from the external medium; intestinal salt and water transport is reduced, and the kidney produces a copious hypotonic urine (see Fig. 1 and Table 1)

It is therefore apparent that salinity adaptation in teleosts will principally involve functional alterations in the gills, intestine and kidneys. The mechanism of these changes can be of at least three kinds.

(1) 'Extracellular' effects, which can be immediate, and important. Examples would be mucus secretion to reduce permeability, or changes in haemodynamics which can profoundly affect branchial or renal function, and can be neurally or hormonally mediated.

(2) Cellular effects, which would reflect changes in enzyme function, such as the induction or the suppression of pumps or permeases, or the modification of metabolism.

Fig. 1. Generalised diagram to illustrate the principal routes for salt and water movement in fish living in seawater (SW) or freshwater (FW).

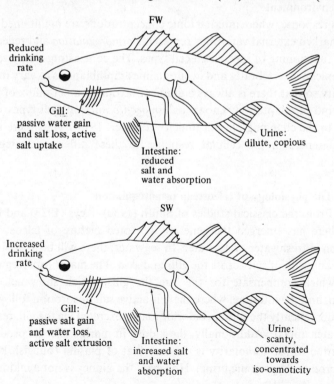

(3) Morphological effects, which include cellular replacement or augmentation, and modification of intercellular junctions, reflecting a change in the structure of the tissue.

Responses at all levels are important, and it is likely that in salt adaptation, as has also been shown for temperature acclimation (for a good example see the review of Smith (1976) on the goldfish), they can be regarded to some extent as sequential, reflecting the time-course of adaptation. Thus in the gill (and see below) an immediate response to the transfer from freshwater to seawater would be mucus secretion and a change in branchial haemodynamics. Subsequently there would be an alteration in the cellular activities of permeases and transport enzymes, and finally the development of new cells

Table 1. *Summary of the changes occurring in the osmoregulatory organs of euryhaline teleosts on adaptation to freshwater or seawater*

Organ	Freshwater adaptation	Seawater adaptation
Gill	Salt absorption (1) Na^+ efflux decreased (2)	Active salt excretion, probably by chloride cells (see Fig. 2)
Oesophagus	Ion permeability low Stratified epithelium	Permeability increases (3) Columnar cells Osmolarity of seawater decreased before it reaches the intestine
Intestine	Low salt transport (4) Decreased mucosal cell Cl^- permeability	Drinking (5). Increased water and NaCl absorption (6); transepithelial potential; ATPase (7)+ junctional permeability (8) increased
Urinary bladder	Decreased P_{osm} in some fish (9,10) Increased NaCl reabsorption ? (11)	High P_{osm} (9, 10)
Kidney	Copious hypotonic urine (1, 12)	Smaller amounts of hypotonic urine (see text for details)

References: 1, e.g. Krogh (1939); 2, Motais & Garcia-Romeu (1972); 3, Yamamoto & Hirano (1978); 4, e.g. Oide & Utida (1967); 5, Hirano (1974); 6, Oide & Utida (1967), Hirano (1967), Skadhauge (1969, 1974); 7, Jampol & Epstein (1970); 8, Ando et al. (1975), Smith et al. (1975); 9, Hirano et al. (1973); 10, Utida et al. (1972); 11, Demarrest (1977), Hirano (1975) but see Forster (1975), Johnson et al. (1972); 12, Smith (1932).

and intercellular connections which would represent the long-term strategy of adaptation.

Strictly, the present brief is to examine cellular adaptation, and so we shall not consider extracellular factors further, although their importance in the gill and kidney may be considerable. Intestine and gills represent the principal organs involved in salt adaptation considered in the cellular context, and we will now review their response in some detail.

One obvious problem lies in defining salt adaptation. The impressive behaviour of some cyprinodont fish to survive any salinity between freshwater and 300% seawater (Lotan, 1971) can be contrasted with the modest ability of the goldfish to survive in 40% seawater following gradual adaptation. In fact there is a distinction in adaptive strategy between these two examples, which emphasises the difference between the response of osmoregulators and osmoconformers, and so we shall consider both.

Interest in the physiology of salt adaptation has been considerable in euryhaline and stenohaline teleosts, and there are many recent reviews and symposia devoted to various aspects of this problem (e.g. Potts & Parry, 1964; Maetz, 1971; Motais & Garcia Romeu, 1972; Maetz & Bornancin, 1975; Kirschner, 1979; Evans, 1980a, b; see also Lahlou, 1980, and articles in *American Journal of Physiology* Symposium (1980), vol. 238, pp. R139–R276).

Regulatory responses to salinity adaptation
Gill

The branchial epithelium is a complicated tissue, in terms both of its structure and of having several different important physiological functions. Aside from its role in osmoregulation it performs respiratory gas exchange, regulates pH (i.e. acid–base balance) and contributes to the excretion of small organic molecules, particularly certain products of nitrogen metabolism. The presence of these multiple functions will complicate most experimental measurements connected with ion transport, so it has been hard to identify exclusively the site and specificity of osmoregulatory changes for salt adaptation in this tissue.

Structurally the gill is designed to have a large surface area, and a vasculature which allows for a measure of control of haemodynamics, consistent with variations in respiratory demand. Fig. 2 illustrates diagrammatically the main features of the branchial epithelium. The epithelial cell layer consists of three cell types: the pavement or respiratory cells, which comprise most of the gill surface, mucus cells, and chloride cells. A fourth cell type, the accessory or adjacent cell, which may represent an immature or degenerate chloride cell, has also been described in seawater fish (Laurent & Dumel, 1980; Hootman & Philpott, 1980). Anatomically the chloride cells are concentrated in the gill primary lamellae, close to the base of the secondary lamellae. The latter consist almost entirely of respiratory cells (Conte, 1969; Laurent & Dumel, 1980; Girard & Payan, 1980). There is a further important difference between the vasculature of the primary and secondary lamellae: the former are venous, whilst the latter are irrigated by the arterioarterial circulation. This distinction allows differential haemodynamic effects which can be important physiologically (Girard & Payan, 1980).

It was shown early on (Krogh, 1939; see also Maetz & Garcia-Romeu, 1964; Evans, 1980b) that in freshwater the gill absorbs Na^+ and Cl^- independently, by two different transport mechanisms. These are now usually identified as a Na^+/NH_4^+ exchange, and a Cl^-/HCO_3^- exchange (although the possibility of Na^+/H^+ and Cl^-/OH^- exchange cannot be ruled out). Sodium chloride absorption by this route is therefore linked to nitrogenous waste metabolism and acid–base balance (including a contribution from the anion-sensitive ATPase: Bornancin, De Renzis & Naon, 1980) and would occur in both freshwater and seawater fish. In freshwater fish it could represent the major route for active salt entry across the gills. Recently this

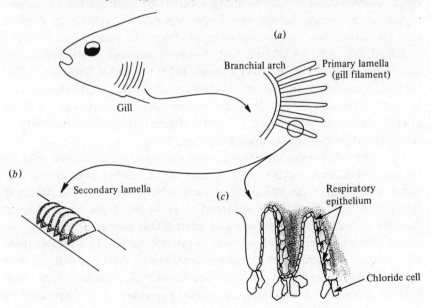

Fig. 2. Fish gills consist of five or six branchial arches. Each branchial arch bears two rows of gill filaments (primary lamellae) which in turn give rise to the plate-like secondary lamellae. (a) Branchial arch; (b) primary lamella; (c) sagittal section through a gill filament showing the position of chloride cells in the primary lamella epithelium at the base of the secondary lamellae.

function has been proposed as being localised in the respiratory cells of the secondary lamellae. Superimposed on these Na^+ and Cl^- movements, and dominant in seawater-adapted fish would be another, regulated sodium chloride excretion which is now thought to occur through the chloride cells.

On adaptation of fish from freshwater to seawater there are striking changes in both the number and morphology of chloride cells (Newstead, 1967; Shirai & Utida, 1970) whilst the respiratory cells do not change their structural characteristics (Sardet, Pisa & Maetz, 1979). The importance of the chloride cells as the principal site for salt excretion associated with adaptation is emphasised by the recent increase in research in this area (e.g. see *American Journal of Physiology* volume for March 1980, the Maetz Memorial Symposium.) Recently, the development of a simple isolated epithelial preparation containing a large number of chloride cells from *Fundulus* opercular cavity skin (Burns & Copeland, 1950; Karnaky & Kinter, 1977) and *Gillichthys mirabilis* jaw skin (Marshall & Bern, 1980) has allowed the mechanism of active Cl^- transport to be followed in some detail.

A key observation on gill response to increased salinity of the medium has been the demonstration of an increase (1.5-to 4-fold) in the activity of $(Na^+ + K^+)$-ATPase levels in microsomal gill homogenates (e.g. Jampol & Epstein, 1970). Although this has now been demonstrated in several species there are reports of no increase being demonstrable in some cases: e.g. European flounder (*Platichthys flesus*) where it has been hypothesised that the enzyme may be cryptic, that is, demonstrable in a cell homogenate but functionally inactive *in vivo*. By using a variety of fractionation techniques including cell separation or membrane separation methods to prepare chloride cells or their membranes, or autoradiography with [^{3}H]ouabain, the increased $(Na^+ + K^+)$-ATPase activity has been localised on the basolateral membranes of the chloride cells (Kamiya, 1972; Sargent & Thomson, 1974; Sargent, Thomson & Bornancin, 1975; Karnaky *et al.*, 1976). The increase in activity as measured by [^{3}H]ouabain binding can occur as early as 1–3 hours after the salinity change (Towle, Gilman & Hempel, 1977; Hossler, 1980) and is influenced by cortisol (Jampol & Epstein, 1970).

There are other important biochemical changes in the chloride cells on seawater adaptation, including an increase in the number of mitochondria, and a deepening of the apical pit of these cells (Bierther, 1970; Maetz & Bornancin, 1975; Karnaky, 1980) (see Fig. 3). In fact, these morphological changes at the apical membrane and intercellular junctions represent the second important functional response in salt adaptation. In seawater there are more cell–cell junctions between adjacent chloride cells, and chloride cells and accessory cells. These junctions are structurally different from those between chloride cells and respiratory cells in showing, as demonstrated by

Fig. 3. Schematic representation of chloride cells in (*a*) freshwater-adapted and (*b*) seawater-adapted teleosts. Changes on adaptation to seawater: (1) deepened apical pit; (2) accessory cell; (3) numerous but very 'leaky' tight junctions; (4) increase in basal infolding, mitochondria and sodium pumps; (5) increased mucopolysaccharide in apical pit; (6) increase in the number of chloride cells. A, accessory cell; C, chloride cell.

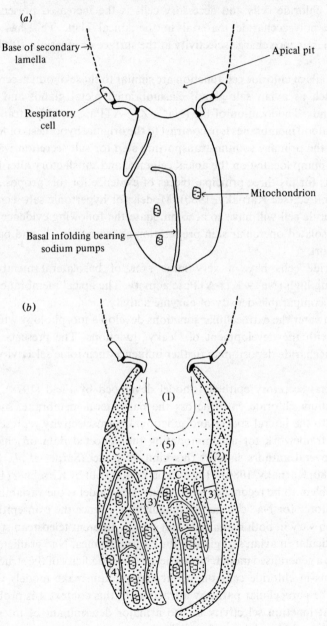

freeze-fracture electron microscopy, very shallow tight junctions made of one strand (Sardet *et al.*, 1979). These kinds of connections represent a minimal junctional link between cells, and are thought to be characteristic of very leaky tight junctions as seen in low-resistance epithelia. An additional feature of the deepened apical pit, and development of contacts between chloride cells or between chloride cells and accessory cells is the increased presence of polyanionic polysaccharide material in the apical cleft. This has been proposed to influence charge selectivity at the surface or junctions (Philpott, 1980).

These details of chloride cell function are similar to those of other secretory epithelia such as avian salt gland, elasmobranch rectal gland, and even salivary glands. The location of the $(Na^+ + K^+)$-ATPase and Na^+ transport in the basolateral membranes is in contrast to the original hypothesis of Maetz (1971) that the primary sodium-transporting step for salt secretion was via the sodium pump located on the apical cell face, and satisfactory alternative explanations for his three principal pieces of evidence for this proposal can now be admitted (see Karnaky, 1980). Models for hypertonic salt secretion via the chloride cell will have to accommodate the following evidence:

(1) The isolated opercular skin preparation shows active Cl^- and passive Na^+ transport.

(2) Chloride cells have a very large area of basolateral membrane, incorporating high $(Na^+ + K^+)$-ATPase activity. The apical membrane does not show a comparable density of enzyme activity.

(3) In seawater the extracellular junctions develop a morphology which is consistent with the development of 'leaky' junctions. The presence of a mucopolysaccharide deposit may further influence their ionic selectivity (see Fig. 3).

The general secretory epithelial model developed by Field (1978), with coupled sodium chloride entry across the basolateral membrane, sodium pumping into the lateral spaces and junctional permselectivity represents a reasonable framework for accommodating the collected data on chloride cells, and several authors support this kind of model (Sardet *et al.*, 1979; Philpott, 1980; Karnaky, 1980). However, as pointed out by Kirschner (1980), the real problem to be reconciled with a universal model is the variability of the driving force for Na^+ exit across the junctions. Since the transepithelial potential can vary in both magnitude and sign in different teleostean species (and in particular in avian salt glands) the electrochemical Na^+ gradient will not easily fit a generalised model. Nevertheless, the specificity of the structural modifications in chloride cells on salinity adaptation make models which emphasise the paracellular pathway attractive. In this context it is probably relevant that junction selectivity is also a major determinant of intestinal transport in euryhaline teleosts (see below).

Before concluding the section on gills it is worth emphasising the direct effects of hormones on the transport properties of chloride cells. Although some of the branchial adaptive changes with salinity may come from direct effects of luminal sodium chloride on chloride cells (Karnaky, 1980) there is good evidence for adrenaline acting via α-adrenergic receptors to reduce Cl^- transport in the opercular skin preparation (Degnan & Zadunaisky, 1978; Zadunaisky & Degnan, 1980; Marshall & Bern, 1980) with an additional β-stimulatory effect (Zadunaisky & Degnan, 1980). Other hormones which modify branchial salt transport include prolactin, urotensin II and cortisol (Marshall & Bern, 1980; Doyle & Epstein, 1972; and see the final section below). It is therefore likely that the rapid, specific intracellular changes in chloride cell structure and enzyme activity on salinity adaptation can be mediated by any or all of the conventional biochemical messenger pathways (Ca^{2+}, cyclic AMP, protein synthesis stimulation via mRNA, etc.) involved in hormone action.

Oesophagus

The permeability of eel oseophagus to ions (Na^+, Cl^- and also K^+, Ca^{2+} and Mg^{2+}) is very low in freshwater and increases significantly after 3 days in seawater, reaching a maximum after 7 days (Kirsch, Guinier & Meens, 1975; Hirano & Mayer-Gostan, 1976; Hirano, 1980). The effect is reversed by returning to freshwater, and involves a histological change where the stratified epithelium seen in freshwater is replaced with a thin lining of simple columnar cells (Yamamoto & Hirano, 1978). The increased oesophageal permeability to salts on salinity adaptation serves to de-salt the ingested hypertonic seawater by increased passive diffusion of sodium chloride, before the fluid enters the stomach and anterior intestine. Conventionally, water will enter the stomach to dilute ingested seawater before absorption of salt and water (Bentley, 1971) and so loss of sodium chloride via the oesophagus removes this gradient for water to enter the stomach.

Fish intestine

Intestine is a major organ involved in the regulatory response to salinity adaptation in euryhaline teleosts. An increased drinking rate in seawater is related to an enhanced absorption of salt and water across the fish anterior intestine, as the ultimate source for replacing body fluid and compensating dehydration by the hypertonic medium. It has been shown from experiments with normal and everted sacs (Oide & Utida, 1967; Hirano, 1967; Lahlou et al., 1975) and perfusions *in vivo* (Skadhauge, 1969, 1974) that net water and salt transport is up to 10-fold higher in seawater eels than freshwater eels or flounders.

When the intestine from seawater teleosts is examined in Ussing chambers

(Huang & Chen, 1971; Hirano & Utida, 1972) it appears, in contrast to other vertebrates, to be dominated by electrogenic Cl^- transport. The transepithelial potential difference (PD) is *serosa* negative and the short-circuit current (I_{sc}) is equivalent to the difference between net Cl^- and net Na^+ transport (Hirano *et al.*, 1975; Ramos & Ellory, 1981). The mechanism of this active Cl^--transporting epithelium has been investigated in some detail by measuring transepithelial Na^+ and Cl^- fluxes, and mucosal uptake following anion or cation substitution (Smith, Ellory & Lahlou, 1975; Field *et al.*, 1978; Ramos & Ellory, 1981). Intracellular Cl^--sensitive microelectrode activity measurements have also been made (Zeuthen, Ramos & Ellory, 1978; Ellory, Ramos & Zeuthen, 1979). Most data are consistent with a model as shown in Fig. 4 (see Field, 1978; Ramos & Ellory, 1981). Mucosal entry of sodium chloride is mediated via a coupled transport system. This is a secondary active transport with the Na^+ gradient producing active intracellular accumulation of Cl^-. The Na^+ is pumped into the lateral intercellular spaces, with Cl^- following down its electrochemical gradient. The apparent electrogenic Cl^- transport occurs as a result of the junctional permselectivity, where the tight junctions are highly cation selective, allowing Na^+ but not Cl^- to diffuse from the lateral spaces back into the lumen; Cl^- preferentially diffuses along the lateral spaces to the serosal side. Although this model is supported by most data, it has recently been questioned (Mackay & Lahlou,

Fig. 4. Model for intestinal salt transport in marine teleosts. Possible sites of regulation in euryhaline teleosts: (1) coupled mucosal sodium chloride entry; (2) sodium pump in basolateral membrane; (3) junctional permselectivity favours sodium. See text for explanation and references. (After Field, 1978.)

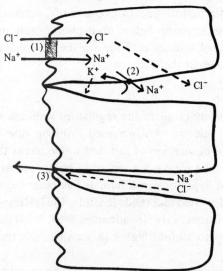

1980). An alternative possibility would be the existence of a chloride pump, possibly related to an anion-sensitive ATPase which would be directly responsible for anion transport (see Hirano *et al.*, 1975; Lahlou *et al.*, 1975).

In terms of salt adaptation, the model suggests three sites where regulation is likely to occur: (1) mucosal entry step; (2) basolateral sodium pump; (3) junctional permselectivity. Although the everted sac experiments, and certainly the transport situation *in vivo* indicate a reduction in net salt transport in freshwater, the Ussing chamber results comparing Na^+ and Cl^- fluxes and electrical measurements in freshwater- and seawater-adapted eels and flounders (Hirano *et al.*, 1975; Smith *et al.*, 1975; Mackay & Lahlou, 1980) do not immediately allow the site of adaptation to be identified. In all studies there is an increased transepithelial PD and I_{sc} in seawater. Net Cl^- transport increases but net Na^+ transport does not (see Table 2 in Mackay & Lahlou, 1980, and Table II-3 in Hirano *et al.*, 1975). However, Jampol & Epstein (1970) have shown an increased $(Na^+ + K^+)$-ATPase activity in seawater-adapted animals, so an increased sodium pump activity seems likely. Since in both eels and flounders the junctional resistance decreases in salt adaptation (Ando, Utida & Nagahama, 1975; Smith *et al.*, 1975), this could account for net Na^+ transport remaining constant while net Cl^- transport increases (the increased Na^+ pumped being lost through increased junctional permeability).

The mucosal sodium chloride entry step is an obvious site for regulation during salinity adaptation. Coupled sodium chloride transport, sensitive to frusemide, has now been widely established as playing a role in epithelial salt transport (Frizzell *et al.*, 1979*a, b*), and Ramos & Ellory (1981) demonstrated that such a mechanism was responsible for salt entry across the luminal border of anterior intestine in American flounder and plaice, respectively. In contrast, Smith *et al.* (1975) and Mackay & Lahlou (1980) failed to demonstrate coupling between Na^+ and Cl^- uptake in intestines from seawater European flounder. However, using Cl^--sensitive microelectrodes, Lau (1982) has recently shown that the intracellular Cl^- activity is higher than electrochemical equilibrium in both seawater- and freshwater-adapted flounders. The mucosal addition of piretanide, a specific inhibitor of the sodium chloride co-transport system (Zeuthen *et al.*, 1978) reduced the intracellular Cl^- activity to the passive equilibrium value in seawater but not freshwater fish. This may indicate a decreased basolateral Cl^- permeability or transport in the freshwater-adapted tissue, so that intracellular Cl^- activity is maintained when the piretanide-sensitive mucosal entry step for sodium chloride is reduced or absent. Thus, in summary, referring to the three major sites for transport regulation in intestine on salinity adaptation, it seems likely that some regulation occurs at all sites.

It may be relevant to consider whether the euryhaline species is normally in freshwater or seawater during its life history. Thus eels, which are considered freshwater-type teleosts, may respond to seawater more dramatically than flounders, which are essentially seawater fish which need to adapt in a complementary way on going into freshwater. An interesting alternative strategy in salt adaptation is shown by the freshwater stenohaline teleosts. For example, goldfish can tolerate 40% seawater provided the transition is gradual (Lahlou, Henderson & Sawyer, 1969). This fish is an osmoconformer and raises its plasma sodium chloride to equal the external salinity. In marked contrast to the euryhaline teleosts, goldfish show a 4-fold *decrease* in intestinal salt and fluid transport on salinity adaptation (Ellory, Lahlou & Smith, 1972). This has been identified as a selective decrease in mucosal Na^+ permeability in this species (Ellory, Nibelle & Smith, 1973). Interestingly there was no change in tissue resistance, indicating that junctional permeability was unchanged. Thus site (1) in Fig. 4 is the principal adaptive route.

Urinary bladder

Although not of major importance in osmoregulation (Skadhauge, 1977) the fish urinary bladder can show adaptive changes on salinity alteration. Lahlou (1967) showed differences in sodium chloride concentration between free-flow urine and bladder urine consistent with sodium chloride reabsorption. This has been identified more completely in transport studies (Lahlou & Fossat, 1971; Hogben *et al.*, 1972; Renfro, 1975, 1977; Fossat & Lahlou, 1979). Hirano *et al.* (1973) studied urinary bladder from a variety of fish species. They found stenohaline and euryhaline fish of freshwater origin had a low osmotic permeability (P_{osm}). Seawater euryhaline teleosts showed a variable P_{osm}, with it decreasing in freshwater, apparently under prolactin control (Utida *et al.*, 1972). Salt reabsorption may also vary. Isolated bladders from *Platichthys stellatus* appeared to show no systematic variation in freshwater and seawater (Johnson *et al.*, 1972; Forster, 1975). However, under short-circuit conditions increased net Na^+ influx was found in freshwater (Demarrest, 1977), and Hirano *et al.* (1973) also found such an increase. Recently, the fish urinary bladder has been studied in terms of its morphology, the two cell types – columnar and cuboidal – being resolved in terms of transport function (Loretz & Bern, 1980). The cells responsible for sodium chloride transport in *Gillichthys mirabilis* were of the columnar type, and showed an increased *electrogenic* Na^+ transport component in seawater, although overall sodium chloride transport was unchanged.

Signals

Finally, we shall consider the pathways by which the fish recognises changes in salinity, and then co-ordinates the responses of the osmoregulatory organs. It is particularly interesting to consider the time-scale of a response which can involve adaptations ranging from short-term regulation by unmasking cryptic proteins, to longer-term replacement of whole cells. The triggering event may be either the direct action of differing external salinities on certain cells, or an indirect signal via changes in internal body fluids. As for most adaptive responses, both neuronal and humoral efferent pathways are involved.

In eels, the best-studied species, there is evidence for an immediate neuronally mediated drinking reflex on transfer to seawater, which is directly dependent on the presence of halide ions (Cl^-) in the medium. Possible receptor sites include the palatal organs, the olfactory system and the lateral line organ (see Hirano, 1974). At present there is no direct evidence for a further role for nerves of either afferent or efferent pathways in salt adaptation, although neuronally mediated haemodynamic effects may be important in both gills and kidney.

Later, adaptive responses following the perturbation of the volume or tonicity of the body fluids will be regulated by pathways similar to those in terrestrial vertebrates (e.g. see Fitzsimons, 1979), integrating the actions of the pituitary, the inter-renal adrenocortical tissues and the renin–angiotensin system.

Hormones

Adrenocortical steroids and prolactin are probably the major hormones concerned with the adaptation of fish to changing salinity. Although their roles are only partially elucidated, and their study beset with experimental difficulties, as a convenient simplification it may be said that cortisol is the principal hormone for seawater adaptation, whilst prolactin plays the major role during transfer to freshwater (Bentley, 1971; Johnson, 1973; Maetz, 1974; Hirano, 1978; Lahlou, 1979).

It is envisaged that cortisol promotes Na^+ transport via stimulation of $(Na^+ + K^+)$-ATPase activity, thus increasing salt excretion by the gill, and salt and water absorption by the intestine, in seawater-adapted fish. Prolactin is thought to limit the permeability to both water and ions, thereby restricting osmotic water gain and diffusional ion loss in freshwater (see Utida *et al.*, 1972). An instructive example instanced by Bentley (1971) is the seasonal migration of the stickleback *Gasterosteus aculeatus*, which moves between seawater and freshwater to breed. Prolactin levels in this fish show seasonal variations consistent with a requirement for this hormone for life in freshwater.

One possible route for prolactin action involves the stimulation of mucus secretion on to the surface of the gill epithelium (e.g. Burden, 1956; Pickford, Pang & Sawyer, 1966). Mineralocorticoids – possibly aldosterone by analogy with its function in higher vertebrates, or low levels of cortisol – are also essential for life in freshwater, acting to promote salt retention in the kidney/bladder complex and salt absorption by the gill.

Conclusions

The present short review attempts to put the various adaptive responses to changing salinity into a cellular context. We have emphasised that three overall strategies are possible in salt adaptation: changing the extracellular medium, e.g. mucus secretion, varying blood flow; changing the intracellular medium, e.g. altering the permeability and transport characteristics of the cell membranes; and replacing the cells with modified cells, which may also be altered at the tissue level in terms of cell–cell connections. (We should also include a fourth strategy of biological importance: behavioural avoidance.)

As mentioned at the outset these strategies are *all* important, and essentially sequential. Our present knowledge is still scanty in many areas of fish physiology, and although recently great progress has been made in understanding the gill, there are still fundamental aspects of salt adaptation in truly euryhaline fish which although impressive are not totally explicable.

References

Ando, M., Utida, S. & Nagahama, M. (1975). Active transport of chloride in the eel intestine with special reference to sea water adaptation. *Comparative Biochemistry and Physiology*, **51A**, 27–32.

Bentley, P. J. (1971). *Endocrines and Osmoregulation*. Berlin, Heidelberg & New York: Springer-Verlag.

Bierther, M. (1970). Die Chloridzellen des Stichlings. *Zeitschrift für Zellforschung und Mikroskopische Anatomie*, **107**, 421–46.

Bornancin, M., De Renzis, G. & Naon, R. (1980). Cl^--HCO_3^--ATPase in gills of rainbow trout: evidence for its microsomal localisation. *American Journal of Physiology*, **238** (*Regulatory Integrative Comparative Physiology* 7), R251–R259.

Burden, C. E. (1956). The failure of hypophysectomised *Fundulus heteroclitus* to survive in fresh water. *Biological Bulletin* (*Woods Hole, Mass.*), **110**, 18–28.

Burns, J. & Copeland, D. E. (1950). Chloride excretion in the head region of *Fundulus heteroclitus*. *Biological Bulletin* (*Woods Hole, Mass.*), **99**, 381–5.

Conte, F. P. (1969). Salt excretion. In *Fish Physiology. Excretion, Ionic Regulation and Metabolism*, vol. 1, ed. W. S. Hoar & D. J. Randall, pp. 241–92. New York: Academic Press.

Degnan, K. J. & Zadunaisky, J. A. (1978). Adrenergic regulation of Cl^-

secretion across the chloride-rich opercular epithelium of the seawater teleost. *Federation Proceedings*, **37**, 652 (abstract).

Demarrest, J. R. (1977). Freshwater acclimatisation in the teleost urinary bladder: changes in transport. *American Zoologist*, **17**, 877 (abstract).

Doyle, W. L. & Epstein, F. H. (1972). Effects of cortisol treatment and osmotic adaptation on the chloride cells of the eel, *Anguilla rostrata*. *Cytobiologie*, **6**, 58–73.

Ellory, J. C., Lahlou, B. & Smith, M. W. (1972). Changes in the intestinal transport of sodium induced by exposure of goldfish to a saline environment. *Journal of Physiology*, **222**, 497–509.

Ellory, J. C., Nibelle, J. & Smith, M. W. (1973). The effect of salt adaptation on the permeability and cation selectivity of the goldfish intestinal epithelium. *Journal of Physiology*, **231**, 105–15.

Ellory, J. C., Ramos, M. & Zeuthen, T. (1979). Cl^- accumulation in the plaice intestinal epithelium. *Journal of Physiology*, **287**, 12P.

Evans, D. H. (1980a). Ionic and osmotic regulation in fish. In *Osmotic and Ionic Regulation in Animals*, ed. G. M. O. Maloiy, pp. 304–90. New York & London: Academic Press.

Evans, D. H. (1980b). Kinetic studies of ion transport by fish gill epithelium. *American Journal of Physiology*, **238**, (*Regulatory Integrative Comparative Physiology 7*), R224–R230.

Field, M. (1978). Some speculations on the coupling between sodium and chloride transport processes in mammalian and teleost intestine. In *Membrane Transport Processes*, vol. 1, ed. J. F. Hoffman, pp. 277–92. New York: Raven Press.

Field, M., Karnaky, K. J., Smith, P. L., Bolton, J. E. & Kinter, W. B. (1978). Ion transport across intestinal mucosa of the winter flounder, *Pseudopleuronectes americanus*. I. Functional and structural properties of cellular and paracellular pathways for Na and Cl. *Journal of Membrane Biology*, **41**, 265–93.

Fitzsimons, J. T. (1979). *The Physiology of Thirst and Sodium Appetite*. *Physiological Society Monograph 35*. Cambridge University Press.

Forster, R. C. (1975). Changes in urinary bladder and kidney function in the starry flounder (*Platichthys stellatus*) in response to prolactin and to freshwater transfer. *General and Comparative Endocrinology*, **27**, 153–61.

Fossat, B. & Lahlou, B. (1979). The mechanism of coupled transport of sodium and chloride in isolated urinary bladder of the trout. *Journal of Physiology*, **294**, 211–22.

Frizzell, R. A., Field, M. & Schultz, S. G. (1979a). Sodium-coupled chloride transport by epithelial tissues. *American Journal of Physiology*, **236**, F1–F8.

Frizzell, R. A., Smith, P. L., Vosburgh, E. & Field, M. (1979b). Coupled NaCl influx across brush border of flounder intestine. *Journal of Membrane Biology*, **46**, 27–39.

Girard, J. P. & Payan, P. (1980). Ion exchanges through respiratory and chloride cells in freshwater and seawater adapted teleosts. *American Journal of Physiology*, **238** (*Regulatory Integrative Comparative Physiology 7*), R260–R268.

Hirano, T. (1967). Effect of hypophysectomy on water transport in isolated intestine of the eel, *Anguilla japonica*. *Proceedings of the Japan Academy*, **43**, 793–6.

Hirano, T. (1974). Some factors regulating water intake by the eel, *Anguilla japonica*. *Journal of Experimental Biology*, **61**, 737–47.

Hirano, T. (1975). Effects of prolactin on osmotic and diffusion permeability of the urinary bladder of the flounder, *Platichthys flesus*. *General and Comparative Endocrinology*, **27**, 88–101.

Hirano, T. (1978). Endocrine control of osmoregulation in fish. In *Comparative Endocrinology*, ed. P. J. Galliard, pp. 209–12. Amsterdam: Elsevier/North-Holland.

Hirano, T. (1980). Effects of cortisol and prolactin on ion permeability of the eel oesophagus. In *Epithelial Transport in the Lower Vertebrates*, ed. B. Lahlou. Cambridge University Press.

Hirano, T., Johnson, D. W., Bern, H. A. & Utida, S. (1973). Studies on water and ion movements in the isolated urinary bladder of selected freshwater, marine and euryhaline teleosts. *Comparative Biochemistry and Physiology*, **45A**, 529–40.

Hirano, T. & Mayer-Gostan, N. (1976). Eel oesophagus as an osmoregulatory organ. *Proceedings of the National Academy of Science*, **73**, 1348–52.

Hirano, T., Morisawa, M., Ando, M. & Utida, S. (1975). Adaptive changes in ion and water transport mechanisms in the eel intestine. In *Intestinal Ion Transport*, ed. J. W. L. Robinson, pp. 301–17. Lancaster, England: MTP Press.

Hirano, T. & Utida, S. (1972). *Seitai no Kagaku*, **23**, 1–16. (In Japanese.)

Hogben, C. A. M., Brandes, J., Danforth, J. & Forster, R. P. (1972). Ion transport by the isolated urinary bladder of a teleost, *Hemitripterus americanus*. *Bulletin of the Mount Desert Island Biological Laboratory*, **12**, 52–5.

Hootman, S. R. & Philpott, C. W. (1980). Accessory cells in teleost branchial epithelium. *American Journal of Physiology*, **238**, (*Regulatory Integrative Comparative Physiology 7*), R199–R206.

Hossler, F. E. (1980). Gill arch of the mullet, *Mugil cephalus*. III. Rate of response to salinity change. *American Journal of Physiology*, **238** (*Regulatory Integrative Comparative Physiology 7*), R160–R164.

Huang, K. C. & Chen, T. S. T. (1971). Ion transport across intestinal mucosa of winter flounder. *Pseudopleuronectes americanus*. *American Journal of Physiology*, **220**, 1734–8.

Jampol, L. M. & Epstein, F. H. (1970). Sodium–potassium-activated adenosine triphosphatase and osmotic regulation by fishes. *American Journal of Physiology*, **218**, 607–11.

Johnson, D. W. (1973). Endocrine control of hydromineral balance in teleosts. *American Zoologist*, **13**, 799–818.

Johnson, D. W., Hirano, T., Bern, H. A. & Conte, F. P. (1972). Hormonal control of water and sodium movements in the urinary bladder of the starry flounder, *Platichthys stellatus*. *General and Comparative Endocrinology*, **19**, 115–29.

Kamiya, M. (1972). NaK-activated ATPase in chloride cells from eel gills. *Comparative Biochemistry and Physiology*, **43B**, 611–17.

Karnaky, K. J. Jr (1980). Ion-secreting epithelia: chloride cells in the head region of *Fundulus heteroclitus*. *American Journal of Physiology*, **238** (*Regulatory Integrative Comparative Physiology 7*), R185–R198.

Karnaky, K. J. Jr, Degnan, K. J. & Zadunaisky, J. A. (1977). Chloride transport across isolated opercular epithelium of killifish: a membrane rich in chloride cells. *Science*, **195**, 203–5.

Karnaky, K. J. Jr, Kinter, L. B., Kinter, W. B. & Stirling, C. E. (1976). Teleost chloride cells. II. Autoradiographic localisation of gill Na–K-ATPase in killifish, *Fundulus heteroclitus*, adapted to low and high salinity environment. *Journal of Cell Biology*, **70**, 157–77.

Karnaky, K. J. Jr & Kinter, W. B. (1977). Killifish opercular skin: a flat epithelium with a high density of chloride cells. *Journal of Experimental Zoology*, **199**, 355–64.

Keys, A. B. (1933). The mechanisms of adaptation to varying salinity in the common eel and the general problem of osmotic regulation in fishes. *Proceedings of the Royal Society of London, Series B*, **112**, 184–99.

Kirsch, R., Guinier, D. & Meens, R. (1975). L'équilibre hydrique de l'Anguille européerine (*Anguilla anguilla*). Etude du rôle de l'oesophage dans l'utilisation de l'eau de boisson et étude de la permeabilité osmotique branchiale. *Journal de Physiologie, Paris*, **70**, 605–26.

Kirschner, L. B. (1979). Control mechanisms in crustaceans and fishes. In *Mechanisms of Osmoregulation in Animals*, ed. R. Gilles, pp. 157–222. New York: Wiley.

Kirschner, L. B. (1980). Comparison of vertebrate salt-excreting organs. *American Journal of Physiology*, **238** (*Regulatory Integrative Comparative Physiology* 7), R219–R223.

Krogh, A. (1939). *Osmotic Regulation in Aquatic Animals*. Cambridge University Press.

Lahlou, B. (1967). Excrétion rénale chez un poisson euryhaline, le flêt (*Platichthys flesus*) caractéristiques de l'urine normale en eau douce et en eau de mer et effets des changements de milieu. *Comparative Biochemistry and Physiology*, **20**, 925–38.

Lahlou, B. (1979). Les hormones dans l'osmoregulation des poissons. In M.A.A.L.I. *Environmental Physiology of Fishes*, pp. 201–40. New York: Plenum Press.

Lahlou, B. (ed.) (1980). *Epithelial Transport in the Lower Vertebrates*. Cambridge University Press.

Lahlou, B., Creresse, D., Benshala-Talet, A. & Porthe-Nibelle, J. (1975). Adaptation de la truite d'élevage à l'eau de mer. Effets sur les concentrations plasmatiques, les échange branchiaux et le transport intestinal du sodium. *Journal de Physiologie, Paris*, **70**, 593–603.

Lahlou, B. & Fossat, B. (1971). Mécanisme du transport de l'eau et du sel à travers la vessie urinaire d'un poisson teleostéen en eau douce, la truite arc en ciel. *Comptes Rendus de l'Académie des Sciences*, **273**, 2108.

Lahlou, B., Henderson, I. W. & Sawyer, W. H. (1969). Sodium exchange in goldfish (*Carassius auratus* L.) adapted to a hypertonic saline solution. *Comparative Biochemistry and Physiology*, **28**, 1427–33.

Lau, K. R. (1982). Active chloride transport in fish intestine. PhD thesis, University of Cambridge.

Laurent, P. & Dumel, S. (1980). Morphology of gill epithelia in fish. *American Journal of Physiology*, **238** (*Regulatory Integrative Comparative Physiology* 7), R147–R159.

Loretz, C. A. & Bern, H. A. (1980). Ion transport by the urinary bladder of the gobiid teleost, *Gillichthys mirabilis*. *American Journal of Physiology*, **239** (*Regulatory Integrative Comparative Physiology* 8), R415–R423.

Lotan, R. (1971). Osmotic adjustments in the euryhaline teleost, *Aphanius dispar* (Rüppell) (Cyprinodontidae). *Zeitschrift für vergleichende Physiologie*, **65**, 455–62.

Mackay, W. C. & Lahlou, B. (1980). Relationships between Na and Cl fluxes in the intestine of the European flounder, *Platichthys flesus*. In *Epithelial Transport in the Lower Vertebrates*, ed. B. Lahlou. Cambridge University Press.

Maetz, J. (1971). Fish gills: mechanisms of salt transfer in freshwater and seawater. *Philosophical Transactions of the Royal Society of London, Series B*, **262**, 209–49.

Maetz, J. (1974). Aspects of adaptation to hypo-osmotic and hyper-osmotic environments. In *Biochemical and Biophysical Perspectives in Marine Biology*, vol. 7, ed. D. C. Malino & J. R. Sargent, pp. 1–167. New York & London: Academic Press.

Maetz, J. & Bornancin, M. (1975). Biochemical and biophysical aspects of salt excretion by chloride cells in teleosts. *Fortschrifte der Zoologie*, **23**, 322–62.

Maetz, J. & Garcia-Romeu, F. (1964). The mechanism of sodium and chloride uptake by the gills of a freshwater fish, *Carassius auratus*. II. Evidence for NH_4/Na and HCO_3/Cl exchanges. *Journal of General Physiology*, **47**, 1209–27.

Marshall, W. S. & Bern, H. A. (1980). Ion transport across the isolated skin of the teleost, *Gillichthys mirabilis*. In *Epithelial Transport in the Lower Vertebrates*, ed. B. Lahlou. Cambridge University Press.

Motais, R. & Garcia-Romeu, F. (1972). Transport mechanisms in the teleostean gill and amphibian skin. *Annual Review of Physiology*, **34**, 141–76.

Newstead, J. (1967). Fine structure of the respiratory lamellae of teleostean gills. *Zeitschrift für Zellforschung und mikroskopische Anatomie*, **79**, 396–428.

Oide, M. & Utida, S. (1967). Changes in water and ion transport in isolated intestines of the eel during salt adaptation and migration. *Marine Biology*, **1**, 102.

Philpott, C. W. (1980). Tubular system membranes of teleost chloride cells: osmotic response and transport sites. *American Journal of Physiology*, **238** (*Regulatory Integrative Comparative Physiology 7*), R171–R184.

Pickford, G. E., Pang, P. K. T. & Sawyer, W. H. (1966). Prolactin and serum osmolality of the hypophysectomised killifish, *Fundulus heteroclitus*, in fresh-water. *Nature, London*, **209**, 1040–3.

Potts, W. T. W. & Parry, G. (1964). *Osmotic and Ionic Regulation in Animals*. Oxford: Pergamon Press.

Prosser, C. L. (1973). *Comparative Animal Physiology*. Philadelphia: W. B. Saunders.

Ramos, M. & Ellory, J. C. (1981). Na and Cl transport across the isolated anterior intestine of the plaice, *Pleuronectes platessa*. *Journal of Experimental Biology*, **90**, 123–42.

Renfro, J. L. (1975). Water and ion transport by the urinary bladder of the teleost, *Pseudopleuronectes americanus*. *American Journal of Physiology*, **228**, 52–61.

Renfro, J. L. (1977). Interdependence of active Na^+ and Cl^- transport by the isolated urinary bladder of the teleost, *Pseudopleuronectes americanus*. *Journal of Experimental Zoology*, **199**, 383–90.

Sardet, C., Pisa, M. & Maetz, J. (1979). The surface epithelium of the teleostean fish gill. Cellular and junctional adaptations of the chloride cells in relation to salt adaptation. *Journal of Cell Biology*, **80**, 96–117.

Sargent, J. R. & Thomson, A. J. (1974). The nature and properties of the inducible sodium-plus-potassium ion dependent ATPase in the gills of the eel (*Anguilla anguilla*) adapted to freshwater and seawater. *Biochemical Journal*, **144**, 69–75.

Sargent, J. R., Thomson, A. J. & Bornancin, M. (1975). Activities and localisation of succinic dehydrogenase and NaK-activated adenosine triphosphatase in the gills of eels (*Anguilla anguilla*). *Comparative Biochemistry and Physiology*, **518**, 75–9.

Shirai, N. & Utida, S. (1970). Development and degeneration of teleostean fish gills. Cellular and junctional adaptations of the chloride cells in relation to salt adaptation. *Journal of Cell Biology*, **80**, 96–117.

Skadhauge, E. (1969). The mechanism of salt and water absorption in the intestine of the eel (*Anguilla anguilla*) adapted to water of various salinities. *Journal of Physiology*, **204**, 135–58.

Skadhauge, E. (1974). Coupling of transmural flows of NaCl and water in the intestine of the eel (*Anguilla anguilla*). *Journal of Experimental Biology*, **60**, 535–46.

Skadhauge, E. (1977). Excretion in lower vertebrates: function of gut, cloaca, and bladder in modifying the composition of urine. *Federation Proceedings*, **36**, 2487–92.

Smith, H. W. (1930). The absorption and excretion of water and salts by marine teleosts. *American Journal of Physiology*, **93**, 480–505.

Smith, H. W. (1932). Water regulation and its evolution in fishes. *Quarterly Review of Biology*, **7**, 1–26.

Smith, M. W. (1976). Temperature adaptation in fish. *Biochemical Society Symposia*, **41**, 43–60.

Smith, M. W., Ellory, J. C. & Lahlou, B. (1975). Sodium and chloride transport by intestine of the European flounder, *Platichthys flesus*, adapted to freshwater or seawater. *Pflüger's Archiv*, **357**, 303–12.

Towle, D. W., Gilman, M. E. & Hempel, J. D. (1977). Rapid modulation of gill Na and K dependent ATPase activity during acclimatisation of the killifish, *Fundulus heteroclitus*, to salinity change. *Journal of Experimental Zoology*, **202**, 179–86.

Utida, S., Hirano, T., Ando, M., Johnson, D. W. & Bern, H. A. (1972). Hormonal control of the intestine and urinary bladder in teleost osmoregulation. *General and Comparative Endocrinology* (*Supplement 3*), 317–27.

Yamamoto, M. & Hirano, T. (1978). Morphological changes in the oesophageal epithelium of the eel, *Anguilla japonica*, during seawater adaptation. *Cell and Tissue Research*, **192**, 25–38.

Zadunaisky, J. A. & Degnan, K. J. (1980). Chloride active transport and osmoregulation. In *Epithelial Transport in the Lower Vertebrates*, ed. B. Lahlou, pp. 185–96. Cambridge University Press.

Zeuthen, T., Ramos, M. & Ellory, J. C. (1978). Inhibition of active chloride transport by piretanide. *Nature, London*, **273**, 678–80.

YUAN LIN

Regulation of the seasonal biosynthesis of antifreeze peptides in cold-adapted fish

A group of glycopeptides and peptides with freezing-point depressing activity has been found in the blood of many cold-adapted marine fish (DeVries, 1980). By some unusual and not well-understood mechanisms, these compounds are capable of lowering the freezing point of the body fluids of these fish and providing them with protection against freezing. Both glycopeptide and peptide antifreeze compounds were found in fish of the Antarctic, Northern Atlantic and Arctic Oceans. Only one type of antifreeze – glycopeptides or peptides – is found in a given species. Similar antifreeze compounds are found in fish of different families.

Antarctic fish inhabit an environment where the annual mean water temperature is -1.86 °C and varies by only 0.2 deg C with season and depth; these fish maintain a constant level (approximately 5% (w/w) of the serum) of antifreeze compound throughout the year. On the other hand, northern fish experience large seasonal temperature variations (-1.7 to 22 °C) in their environment over the course of the year and as a result there is a large seasonal fluctuation in the concentrations of antifreeze compounds in the blood (Petzel, Reisman & DeVries, 1980). Very little is known about how changing environmental factors lead to changes in concentrations of the antifreeze compounds in northern fish. However, an understanding of the underlying control mechanism of antifreeze production may give some insight into gene regulation to meet the demands of a seasonally fluctuating environment.

Chemical structure and mode of action of antifreeze compounds

Both the glycopeptide and peptide antifreezes have been purified from blood of different fish. In each case a family of antifreeze compounds similar in composition but different in size has been found. Fig. 1 illustrates the chemical structure of a glycopeptide antifreeze found in an Antarctic fish, *Trematomus borchgrevinki*, and that of a peptide antifreeze found in winter flounder, *Pseudopleuronectes americanus*. At a first glance they do not seem to bear any resemblance to each other except for the high content of alanine ($\sim 66\%$). However, each molecule possesses a repeating unit of -(Thr-

Ala-Ala)- for glycopeptides and a longer repeating unit of -(Thr-X-X-polar amino acid-X-X-X-X-X-X)- for the winter flounder antifreeze (DeVries & Lin, 1977; Lin & Gross, 1981). If one takes a closer look, sections of -(Thr-Ala-Ala)- can also be found in the peptide antifreeze molecule. Whether such a similarity has any significance in the genetic evolution of these compounds is not known.

How do antifreeze compounds lower the freezing point of an aqueous solution? Dr Arthur DeVries discovered that these compounds in an aqueous solution lower the freezing point but have little effect on the melting point. The effect of freezing-point depression reaches maximum at a concentration of 20 mg ml^{-1}. At this concentration the freezing point is depressed by about 1 deg C i.e. from -1 °C to -2 °C (DeVries, 1971). On a molar basis the antifreeze depresses the freezing point 200 times more effectively than sodium chloride. The mechanism of this non-colligative freezing-point depression has been investigated and a model proposed to account for this unusual phenomenon (Raymond & DeVries, 1977; DeVries & Lin, 1977). The model suggests a recognition of the ice by the glycopeptides and peptides, and their binding to it, which results in inhibition of crystal growth. The adsorbed antifreeze molecules appear to present a barrier to the advancing front of water molecules which are joining the ice lattice during crystal growth. Inhibition of growth apparently results from the fact that the advancing

Fig. 1. Structure of (a) a glycopeptide antifreeze from an Antarctic fish and (b) a peptide antifreeze from winter flounder. Different sizes of glycoprotein antifreeze contain different numbers of the repeating unit -Thr-Ala-Ala-; occasionally an alanine residue is replaced by a proline. Disacch., galactosyl-N-acetylgalactosamine. (From DeVries & Lin, 1977.)

(a)

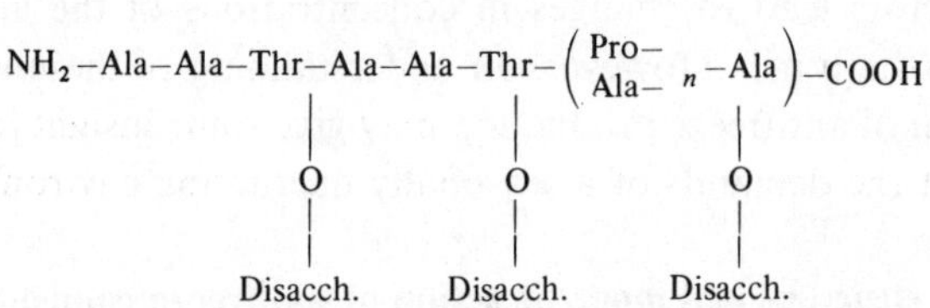

$$NH_2-Ala-Ala-Thr-Ala-Ala-Thr-\left(\begin{matrix}Pro-\\Ala-\end{matrix}\right)_n-Ala\right)-COOH$$

| | |
O   O   O
| | |
Disacch. Disacch. Disacch.

(b)

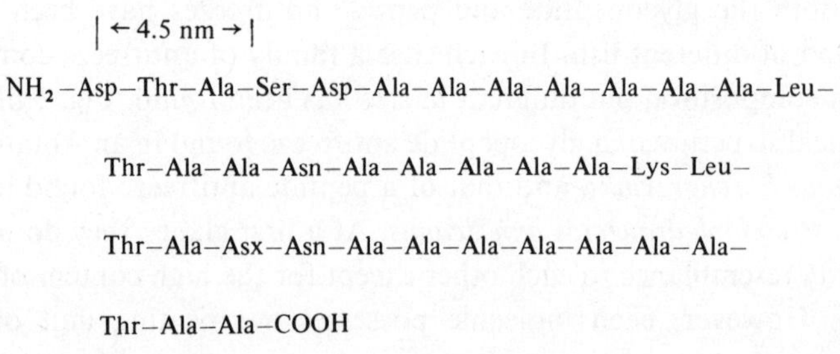

| ← 4.5 nm → |

NH$_2$–Asp–Thr–Ala–Ser–Asp–Ala–Ala–Ala–Ala–Ala–Ala–Ala–Leu–

Thr–Ala–Ala–Asn–Ala–Ala–Ala–Ala–Ala–Lys–Leu–

Thr–Ala–Asx–Asn–Ala–Ala–Ala–Ala–Ala–Ala–Ala–

Thr–Ala–Ala–COOH

ice-front can neither 'overgrow' the adsorbed antifreeze molecules nor pass in between them.

In the case of the glycopeptide antifreezes, chemical modification studies have shown that a specific structure is necessary for the non-colligative freezing-point depression as well as for them to adsorb to ice. The mechanism by which the antifreeze molecules bind to the ice lattice has not been elucidated, but the presence of large numbers of hydroxyl groups on the saccharide side chains suggests to us that hydrogen bonding is involved in the binding process (DeVries, 1980). If this were so, then one would expect to find a repeat spacing of the hydroxyl groups of adjacent saccharides that corresponds to a repeat spacing of the oxygens in the ice lattice. Discovering such a relationship would depend upon knowledge of the secondary structure of the glycopeptides which at this time has not been determined. With the peptide antifreeze, however, recent circular dichroism studies have shown that most of the molecule is in the form of an α-helix, and viscosity studies indicate that it is rod-shaped (DeVries & Lin, 1977). A molecular model with α-helical configuration has been constructed for an antifreeze peptide. In this model all polar side chains are co-planar and separated by 4.5 nm, the distance between neighbouring oxygen atoms in the ice lattice. The structure also aligns all the non-polar side chains on the opposite side of the molecule. From this model it is conceivable that antifreeze peptides can bind to the surface of ice crystals through hydrogen bonding, while the non-polar residues form a barrier preventing water molecules from joining the ice lattice. Such an adsorption–inhibition theory was proposed as a mechanism by which these peptides lower the freezing point of an aqueous solution (DeVries & Lin, 1977). It is presently unknown whether such a molecular structure exists in all antifreeze peptides.

Seasonal variation of antifreeze concentration in winter flounder

Winter flounder (*Pseudopleuronectes americanus*) is found in the shallow waters off the Atlantic coast of North America. In distribution it ranges as far south as Georgia and as far north as Labrador. Only serum from the northern populations is enriched with antifreeze peptides. Seasonal variations in the concentration of antifreeze peptides occur, with highest concentrations in the winter months. Fig. 2 illustrates the water temperature, photoperiod, and antifreeze concentration in the winter flounder collected at different months during 1977–8 in Shinnecock Bay, Long Island, New York. There is a dramatic increase in antifreeze concentration in late autumn and a decrease in late spring. The highest concentration coincides with the lowest water temperature and shortest daylength of the year (Petzel *et al.*, 1980). Laboratory acclimation studies have indicated that both temperature and photoperiod control the synthesis and degradation of antifreeze peptides, i.e. low temperature and short day induce synthesis while high temperature and long day induce degradation (Duman & DeVries, 1974).

Regulation of antifreeze biosynthesis in winter flounder

In order to understand the cellular events which lead to the biosynthesis of antifreeze peptides under the influence of environmental factors, attempts were made to isolate messenger RNA (mRNA) specific for the synthesis of the peptides (Lin, 1979; Lin & Long, 1980). Like many other serum proteins, antifreeze peptides of winter flounder are synthesised in the liver. Livers collected from flounder at different times of the year were lyophilised, ground to a fine powder and extracted with phenol and chloroform. The RNA was precipitated with ethanol and passed through an oligo-d(T)-cellulose affinity column which selectively binds to the poly(A)-containing mRNA. The mRNA was eluted from the column and translated into proteins by means of cell-free protein synthetic systems such as wheat germ lysate and rabbit reticulocyte lysate in the presence of radioactive amino acids. These cell-free protein synthetic systems faithfully synthesise proteins according to the exogenous mRNA added. The radioactive proteins can then be analysed by denaturing polyacrylamide gel electrophoresis. Fig. 3 is an autoradiogram showing proteins made from mRNA isolated from livers of flounder caught

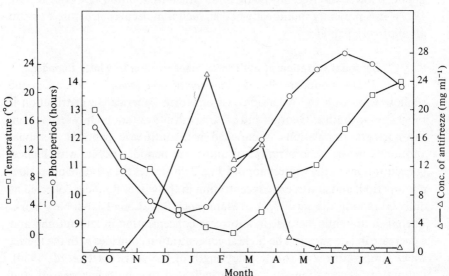

Fig. 2. Temperature and photoperiod in Shinnecock Bay, Long Island, New York during 1977–8, and the antifreeze concentration in the serum of winter flounder collected there during the same period. (From Petzel *et al.*, 1980.)

in different months. The dark band indicated by the arrow is the antifreeze peptide which can be precipitated by antibody prepared from purified winter flounder antifreeze peptide (Fig. 4).

During winter and spring (November to April) few proteins other than antifreeze peptide are made. In the summer no antifreeze mRNA can be detected, judging from the cell-free synthesis products. The appearance and disappearance of this mRNA coincides with changing serum concentration of the antifreeze peptide. These observations suggest that the control of antifreeze biosynthesis resides mostly in the transcriptional or post-transcriptional processing and transportation of antifreeze mRNA (Lin & Long, 1980).

Molecular cloning of DNA complementary to antifreeze mRNA

In order to study the structure, organisation, and regulation of expression of the antifreeze gene, it was necessary to isolate or synthesise a pure DNA 'probe' which could be used to hybridise with the antifreeze gene. Due to the small amount of antifreeze mRNA that exists in the liver, and the tedious procedures involved in the mRNA purification, it is extremely difficult to obtain large enough quantities of a single species of antifreeze

Fig. 3. An autoradiogram of proteins synthesised by wheat germ lysate with mRNA from winter flounder collected at different months of the year. [^{3}H]leucine was added to the reaction mixture. C is a control in which no mRNA was added. In A, S, O, etc., mRNA from the months of August, September, October, etc., was added. AF, antifreeze peptide.

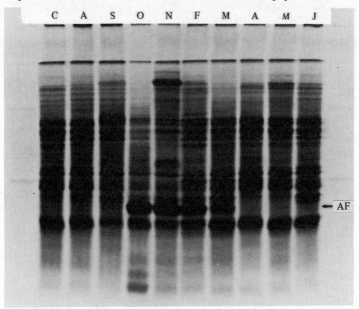

mRNA to be used as a probe for hybridisation analysis. Fortunately, many recently developed techniques in recombinant DNA research have made it possible to clone a double-stranded complementary DNA (cDNA) from a partially purified mRNA and to identify those clones which contain the DNA sequence of interest. The DNA sequence can subsequently be isolated and used as a probe for hybridisation analysis. The fact that antifreeze mRNA contains only 450 nucleotides in length and that it has been purified to more than 50% in purity makes antifreeze DNA much easier to clone and screen than other, more complicated and less pure mRNAs.

The mRNA purified from the winter flounder liver was used as a template for the synthesis of single-stranded and then double-stranded cDNA by reverse transcriptase. This synthetic antifreeze gene was inserted into a bacterial plasmid. (Plasmids are small circular DNA that often contribute to the antibiotic resistance of the bacteria.) The plasmids containing the antifreeze gene are taken up by and propagated in *E. coli* (Lin & Gross, 1981). *E. coli* containing antifreeze gene were identified by their loss of resistance to antibiotics and ampicillin. Since the cDNA insertion was made at the site where the ampicillin-resistant gene is located, those *E. coli* with insertions should not grow in ampicillin plates. The ampicillin-sensitive clones were used to screen for plasmids containing antifreeze cDNA by the hybridisation selection technique. This is based on the principle that if a hybrid plasmid

Fig. 4. Fluorogram of wheat germ cell-free protein synthetic products before (1) and after (2) immunoprecipitation with rabbit anti-antifreeze peptides. Proteins were labelled with [³H]leucine. (From Lin & Long (1980). © 1980 American Chemical Society.)

contains the antifreeze DNA it will hybridise with the antifreeze mRNA which can be isolated and translated into antifreeze peptide in a cell-free protein synthetic system. Plasmids containing antifreeze cDNA were lysed from *E. coli* and recovered by caesium chloride banding centrifugation. The cDNA insert can be recovered by restriction endonucleases.

One of the cDNA isolated in this manner is 380 nucleotides in length and contains the entire coding sequence for an antifreeze peptide including the leader sequence. The nucleotide sequence and the predicted amino acid sequence are illustrated in Fig. 5. Some of the most striking features in this cDNA are:

Fig. 5. Nucleotide sequence of an antifreeze cDNA isolated from a plasmid. Numbers above the amino acids refer to positions of amino acids in the peptide. Negative numbers refer to the leader sequence. Numbers below refer to the positions of nucleotides in the cDNA sequence. Three segments of repeating sequence are enclosed in parentheses. *Hin*fl and *Sau*3Al are sites for restriction enzymes. (From Lin & Gross, 1981.)

TC 1	ACT	TTT	CAC	TGT	CGA	ACA	ATT	GAT	TTC	TTA	TTT	TGA
			−21 Met ATG 39	Arg AGA *Hin*fl	Ile ATC	Thr ACT	Glu GAA	Ala GCC	Asn AAC	Pro CCC	Asp GAC	Pro CCC
	Asp GAC 72	Ala GCC	Lys AAA	Ala GCC	Val GTC	Pro CCT	Ala GCC	Ala GCA	Ala GCC	Ala GCC	−1 Pro CCA
	1 Asp GAC 105	2 Thr ACC	3 Ala GCC	4 Ser TCT	5 Asp GAT	6 Ala GCC	7 Ala GCA	8 Ala GCA	9 Ala GCA	10 Ala (GCC	11 Ala GCC	12 Ala GCC
		13 Thr ACC 141	Ala GCA	Ala GCC	Thr ACC	Ala GCC	Ala GCC)	Ala GCC	Ala GCA	Ala (GCA	Ala GCC	Ala GCC
		24 Thr ACC 174	Ala GCA	Ala GCC	Thr ACC	Ala GCC	Ala GCC)	Ala GCA	Ala GCA	Ala (GCC	Ala GCC	Ala GCC
		35 Thr ACC 207	Ala GCA	Ala GCC	Thr ACC	Ala GCC	Ala GCC)	Lys AAA	Ala GCC	Ala GCA	Ala GCC	Leu CTA
		46 Thr ACC 240	Ala GCC	Ala GCC	Asn AAC	Ala GCC	Ala GCC	Ala GCC	Ala GCC	Ala GCA	Ala GCA	Ala GCC
		57 Thr ACC 273	Ala GCC	Ala GCC	Ala GCA	Ala GCC	Ala GCC	Arg AGA	GGT	TAA	*Sau*3Al ↓ GGA	TCC

(1) The periodical occurrence of the trinucleotide ACC coding for threonine, which appears at every eleventh position. The same periodicity is found in all winter flounder antifreeze peptide sequences.

(2) The presence of long stretches of trinucleotides coding for alanine residues. Even though there are four codons for alanine, only GCC and GCA are present in the nucleotide sequence coding for the antifreeze peptide. (Uneven usage of codons is observed in many other eukaryotic mRNAs.) The coding sequence thus contains approximately 80% G+C.

(3) The presence of three segments of repeating sequences 27 bases in length in the coding region. A regular pattern of -(Thr-X-X-polar amino-acid-X-X-X-X-X-X)- is observed throughout the entire sequence. This basic repeating sequence and its spatial arrangement are important for the binding of the antifreeze molecule to the ice crystal and the retardation of the growth of a seed crystal. Except for the leader sequence, leucine, lysine and arginine are found only in the last portion of the amino acid sequence coded by the cDNA. Different numbers of leucine, lysine and arginine residues are found in all the antifreeze peptides even though these are probably not directly involved in the antifreeze activity, i.e. binding of the peptide to the ice crystal. It is possible that there exists a family of winter flounder antifreeze peptides with amino acid substitutions at the neutral region (-X-X-X-X-) of the molecule. As long as the distance between threonine and the polar amino acid is maintained, many of the amino acids at the 'non-strategic' positions can be replaced by others.

Now that the cDNA has been cloned into *E. coli* it is possible to produce large quantities of pure antifreeze cDNA. These cDNA molecules can be made radioactive by enzymatic reactions using polymerase I and the resulting radioactive cDNA probe can then be used to characterise the seasonal regulation of gene activity in winter flounder as follows:

(1) *The number of antifreeze genes in the genome.* Winter flounders collected from different geographical locations have been shown to produce different amounts of antifreeze peptides in the winter. Do northern populations have larger numbers of antifreeze genes than southern populations? Do low temperature and short photoperiod cause gene amplification? Such questions can be answered by hybridising the radioactive cDNA probe to the total genomic DNA after this has been cleaved into fragments with endonucleases. The fragments can be separated by gel electrophoresis and the number of fragments which hybridise with the radioactive cDNA probe gives an indication of the number of antifreeze genes that exists in the genome.

(2) *The structure and organisation of antifreeze genes.* Do these genes contain intervening sequences (i.e. non-coding regions)? Since the cDNA probe is originally synthesised from the mature cellular mRNA it contains

no intervening sequence. An imperfect match of the cDNA to the antifreeze gene will indicate the presence of the intervening sequences which might play important roles in the regulation of gene expression.

(3) *Primary structure of cDNA.* Techniques are available for the rapid sequencing of DNA. Sequencing of cDNA reveals the primary structure of mRNA with its coding sequence and its flanking sequences which are important for the translational activity of the mRNA.

(4) *The regulation of expression of the antifreeze gene by temperature and photoperiod.* mRNA concentration can be measured directly by its hybridisation to the radioactive cDNA probe. The quantity of mRNA transcribed by the antifreeze gene as a result of acclimation or with the progression of seasons can thus be measured.

The unique molecular structure of winter flounder antifreeze peptides presents an opportunity to study how repetitive DNA sequences are duplicated and conserved during the course of evolution. A radioactive cDNA probe can be prepared and used to search for and identify the antifreeze gene in the genome and, more importantly in the present context, to quantitate the gene product in response to changes in temperature and photoperiod. Ultimately, it is hoped that these studies will lead to an understanding of how the transcription of antifreeze genes is regulated, which may have general significance in cellular mechanisms of acclimatisation.

Concluding remarks

Ever since the antifreeze compounds of marine fish were discovered by Dr Arthur L. DeVries more than 15 years ago they have generated a great deal of interest in scientists of different disciplines. Researchers are fascinated by these simple molecules which modify the physical and chemical properties of the blood and enable the fish that possess them to invade habitats of sub-zero water temperatures. Physical chemists are trying to understand how they modify the structure of water or ice in order to depress the freezing point but not the melting point. Physiologists are dissecting the fine structure of the kidneys in these fish to understand how they can conserve high concentrations of antifreeze peptides in their blood and prevent them being filtered into the urine. Molecular biologists are probing into the antifreeze genes to study their structure and regulation. Such studies will undoubtedly broaden our knowledge of the molecular mechanisms of cellular adaptation as well as our understanding of how new genes are evolved or pre-existing genes are turned on to produce a group of new proteins which are needed by the organism to survive in changing environments.

Research sponsored in part by the National Cancer Institute, DHHS, under Contract no. N01-C0-75380 with Litton Bionetics, Inc. The contents of this publication do not necessarily reflect the views or policies of the Department of Health and Human Services, nor does mention of trade names, commercial products, or organisations imply endorsement by the US Government.

References

DeVries, A. L. (1971). Glycoproteins as biological antifreeze agents in Antarctic fishes. *Science*, **172**, 1152–5.

DeVries, A. L. (1980). Biological antifreezes and survival in freezing environments. In *Animals and Environmental Fitness*, ed. R. Gilles, pp. 538–607. Oxford & New York: Pergamon Press.

DeVries, A. L. & Lin, Y. (1977). Structure of peptide antifreeze and mechanism of adsorption to ice. *Biochimica et biophysica Acta*, **495**, 388–92.

Duman, J. G. & DeVries, A. L. (1974). The effect of temperature and photoperiod on the antifreeze production of cold water fishes. *Journal of Experimental Zoology*, **190**, 89–98.

Lin, Y. (1979). Environmental regulation of gene expression: *in vitro* translation of winter flounder antifreeze mRNA. *Journal of Biological Chemistry*, **254**, 1422–6.

Lin, Y. & Gross, J. G. (1981). Molecular cloning and characterization of winter flounder antifreeze cDNA. *Proceedings of the National Academy of Sciences, USA*, **78**, 2825–9.

Lin, Y. & Long, D. J. (1980). Purification and characterization of winter flounder antifreeze peptide mRNA. *Biochemistry*, **19**, 1111–16.

Petzel, D. H., Reisman, H. M. & DeVries, A. L. (1980). Seasonal variation of the peptide in winter flounder. *Journal of Experimental Zoology*, **211**, 63–9.

Raymond, J. A. & DeVries, A. L. (1977). Adsorption inhibition as a mechanism of freezing resistance in polar fishes. *Proceedings of the National Academy of Sciences, USA*, **74**, 2589–93.

DENNIS A. POWERS

Adaptation of erythrocyte function during changes in environmental oxygen and temperature

Comparative biochemists and physiologists have provided ample evidence that poikilotherms are, in general, adapted to the environments in which they live (reviewed by Prosser, 1973). Thus, it is not surprising that the blood–oxygen affinities of fishes are compatible with the oxygen concentrations of their habitats (Powers *et al.*, 1979*a*; Powers, 1980). In general, fish living in low oxygen environments have high oxygen affinities while those living in high oxygen environments have lower affinities. Moreover, fish living in environments that have periodic changes in oxygen possess the necessary physiological plasticity to maintain respiratory homeostasis. Maintenance of homeostasis is achieved by a host of molecular, cellular, genetic and physiological mechanisms. We shall discuss some of those mechanisms of adaptation.

Haemoglobin

The primary function of haemoglobin is to carry oxygen from the 'organism–environment interface' (e.g. the gills, lungs) to the respiring tissues. This physiological function depends on the ability of the haemoglobin to form a reversible complex between oxygen and the ferrous iron in haematoporphyrin. Under constant physical conditions and in the absence of modifying molecules the affinity of the haemoglobin for oxygen depends primarily on the hydrophobic nature of the haem pocket and on the specific roles of amino acid residues that directly or indirectly affect the oxygen affinity of the macromolecule. Since these amino acids are coded by the globin genes, the intrinsic oxygen–haemoglobin affinity is genetically determined.

One strategy for adapting to a changing environment is to have a number of genetically distinct haemoglobins some of which would have unique functional characteristics. In contrast to the situation in most mammals and birds, the presence of multiple haemoglobins is common among fish species (reviewed by Riggs, 1970). While some fish have unique haemoglobins that help them adapt to specific habitats (e.g. Powers, 1972; Powers & Edmundson, 1972), the vast majority of multiple haemoglobins within species do not appear to serve unique functions (Powers, 1980). That is to say, many fish have several haemoglobins coded for by different genes but the physiological functions of these isohaemoglobins are usually (but not always) indistinguishable. This type of adaptation has been extensively reviewed elsewhere (Riggs, 1970; Brunori, 1975; Bonaventura, Bonaventura & Sullivan, 1975; Powers, 1980).

Interrelationships between haemoglobin and ligands

Since the ability of haemoglobin to bind oxygen is affected by the presence of protons and organic phosphates, considerable physiological plasticity can be provided by the regulation of the intracellular concentrations of these allosteric modifiers. However, these effects are not independent. For example, the binding of protons to haemoglobin not only reduces the oxygen–haemoglobin affinity (i.e. the Bohr and Root effects: see Root, 1931; Riggs, 1971) but also increases the affinity of haemoglobin for organic phosphates such as ATP. Moreover, the binding of organic phosphates to haemoglobin reduces oxygen–haemoglobin affinity thus amplifying the Bohr effect. This interdependence between binding affinities is referred to as linked thermodynamic functions.

The thermodynamic linkage relationships between the binding of protons, organic phosphate and oxygen to haemoglobin are well-documented (e.g. see Wyman, 1964; Riggs, 1970; Brunori, 1975; Greaney, Hobish & Powers, 1980) and are described mathematically by the theory of linked functions which was developed by Wyman (1964).

When haemoglobin (M) binds four molecules of a ligand such as oxygen (X) and one molecule of a second ligand (D) such as organic phosphate, the equilibria can be described by

$$M \longleftrightarrow MX \longleftrightarrow MX_2 \longleftrightarrow MX_3 \longleftrightarrow MX_4$$
$$\updownarrow \quad \updownarrow \quad \updownarrow \quad \updownarrow \quad \updownarrow$$
$$MD \longleftrightarrow MDX \longleftrightarrow MDX_2 \longleftrightarrow MDX_3 \longleftrightarrow MDX_4$$

If one assumes that temperature, pH, etc., are held constant, this can be simplified to:

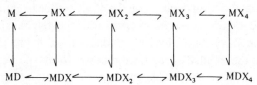

where the association constants are symbolised by Ks, and the super-prescripts indicate the degree of oxygenation (0, x and 4 are for non-, intermediate and fully oxygenated haemoglobin, respectively).

When the concentration of a third ligand, such as protons, is allowed to vary, the system becomes even more complex. For example, the binding of protons (H) and organic phosphate (D) to deoxyhaemoglobin is depicted as:

$$M \rightleftharpoons MH \rightleftharpoons MH_2 \rightleftharpoons \cdots MH_n \rightleftharpoons MH_{n+m} \rightleftharpoons \cdots MH_\infty$$
$$\updownarrow \quad \updownarrow \quad \updownarrow \quad \updownarrow \quad \updownarrow \quad \updownarrow$$
$$MD \rightleftharpoons MHD \rightleftharpoons MH_2D \rightleftharpoons \cdots MH_nD \rightleftharpoons \cdots MH_{n+z}D \rightleftharpoons \cdots MH_\infty D$$

where the subscripts n, m and z refer to the number of protons bound per deoxyhaemoglobin tetramer.

A similar scheme could be devised for each of the various oxygenated haemoglobin species, to give a complex three-dimensional reaction scheme.

At low proton concentration (i.e. pH 10 or greater), the effects of variation in the number of bound protons and organic phosphate molecules upon the binding of oxygen are minimal. When there are n protons bound to the haemoglobin, then:

$$\frac{[MX_4H_nD]}{[MX_4H_n]} \cong \frac{[MH_nD]}{[MH_n]}$$

If a series of organic phosphate binding experiments is performed on deoxyhaemoglobin, starting at the 'minimal' proton concentration (i.e. when n protons are bound) and systematically lowering the pH until the organic phosphate–haemoglobin affinity is maximised, then this pH region can be described by:

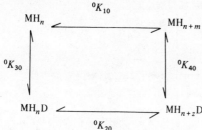

where $^0K_{10}$, $^0K_{20}$, $^0K_{30}$ and $^0K_{40}$ are the appropriate association equilibrium constants and the subscripts (n, m and z) indicate the number of protons bound per deoxyhaemoglobin tetramer. If, within this defined pH range, another ligand such as oxygen is introduced, the overall simultaneous equilibria can be represented by the complex three-dimensional diagram in Fig. 1.

Examination of Fig. 1 indicates a number of possible ways that the oxygen–haemoglobin affinity may be regulated. For example, it can be differentially affected by changing intracellular concentrations of ligands (e.g. organic phosphates), by modifying one or more of the ligand–protein binding constants, or by a combination of these. The dissociation of all of these complexes is increased by increasing temperature; this provides another possible means of adaptation in relation to environmental temperature.

Organic phosphates

The major organic phosphate in fish erythrocytes is either adenosine triphosphate (ATP) (e.g. Gillen & Riggs, 1971; Powers & Edmundson, 1972; Wood & Johansen, 1972) or guanosine triphosphate (GTP) (Geohegan & Poluhowich, 1974), while in mammalian erythrocytes it is 2,3-diphosphoglycerate (2,3-DPG) (Benesch & Benesch, 1967). The effects of ATP and GTP on the oxygen-binding properties of fish haemoglobins are similar to those of 2,3-DPG on mammalian haemoglobins, in that they decrease the affinity of haemoglobin for oxygen.

ATP is a highly charged anion over the physiologically important pH range (i.e. pH 6.9 to 8.0). Three of the four protons are fully dissociated while the fourth is partially dissociated. At pH 8.0 approximately 97% of the ATP is

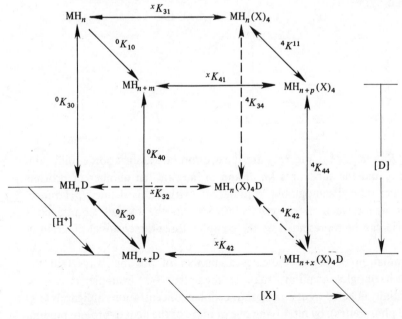

Fig. 1. A generalised scheme of the linkage equilibria which relate the binding of protons (H), organic phosphate (D) and oxygen (X) to haemoglobin tetramers (M). The subscripts (n, z, p, m, x) indicate the number of protons bound. See text for explanation of K values. Haemoglobin with oxygen bound is represented with all four sites saturated. The arrows at the edges of the diagram indicate increasing concentration of appropriate ligand (H, X and D).

in the fully ionised state (ATP^{4-}), while at pH 6.9 only 71% of the ATP is fully ionised, the remaining molecules having a proton bound to the terminal phosphate (ATP^{3-}). The pK values for the other ionisable phosphates are so low that they are well outside the pH range and can thus be ignored. It is reasonable to assume that the fully (ATP^{4-}) and the partially (ATP^{3-}) ionised states would have different affinities for haemoglobin.

Since ATP is the major organic phosphate modifier of many fish haemoglobins, the thermodynamics of ATP binding to fish haemoglobins have been investigated. Greaney *et al.* (1980) have shown that the affinity of haemoglobin for the fully ionised species of ATP is approximately six orders of magnitude greater than that for the partially ionised form. Therefore, intra-erythrocyte pH has a profound effect on the binding of ATP and, hence, on oxygen–haemoglobin affinity. In addition to its allosteric effect on oxygen affinity, organic phosphate has an influence on the Donnan distribution of protons across the erythrocyte membrane, thereby affecting intra-erythrocyte pH (Wood & Johansen, 1972).

Adaptation to temperature changes

The oxygen requirements of fish are correlated with temperature. For example, Fry & Hart (1948) showed that oxygen uptake by goldfish is increased by about 250% over a 10 deg C range. As temperature increases, the availability of oxygen in water decreases because its solubility decreases with increasing temperature (Henry's Law), and because elevated biological activity (e.g. bacterial and planktonic) at high temperatures often reduces oxygen to below saturation. Consequently, at higher temperatures fish require more oxygen but less is available. Fish respond by a series of mechanisms (Powers, 1974; Powers *et al.*, 1979*b*; Powers, 1980; Houston, 1980; Brunori, 1975).

It is clear that temperature affects a wide variety of biochemical phenomena that can either directly or indirectly influence oxygen equilibria. The overall enthalpy of blood–oxygen affinity, ΔH, represents both the intrinsic heat of oxygen binding and contributions due to other ligand-linked processes. It has been shown that while a selective advantage for a thermally insensitive haemoglobin can be envisaged for some species, arguments for a generalised evolutionary strategy for reduced thermal sensitivity are not compelling (Powers, 1980). Thus if evolution has favoured a decrease in the temperature sensitivity of oxygen–haemoglobin affinity, it must be primarily associated with the regulation of intracellular pH, the levels and types of organic phosphates, and other ligand-linked phenomena, rather than with the selection of particular haemoglobin isoforms *per se*. Thus, thermal adaptation of blood–oxygen affinity must be primarily at the level of the whole erythrocyte rather than an intrinsic property of oxygen binding to genetically distinct haemoglobins.

In the presence of air-saturated water, the blood–oxygen affinity of acclimated fish (*Fundulus heteroclitus*) showed either perfect thermal compensation (Fig. 2) or slight overcompensation, depending on the time of year. Adaptation was achieved, in part, by reducing intra-erythrocyte organic phosphate (Fig. 3). However, the intra-erythrocyte ATP levels were significantly higher than in a previous experiment (e.g. Powers & Powers, 1975) and there was no change in haematocrit (Greaney & Powers, 1979). These differences, along with the results of studies of hypoxia (Greaney & Powers, 1978) have indicated that thermal acclimation may involve both hypoxic and thermal effects, that may act either independently or, perhaps, synergistically.

The effect of temperature on ATP concentrations

Since a physical dependence exists between oxygen solubility and water temperature (oxygen concentration is 12.5 p.p.m. at 10 °C compared with 7.5 p.p.m. at 30 °C), our observation that *F. heteroclitus* lowers erythrocyte ATP levels when acclimated to elevated temperatures (group I, Table 1) prompted us to ask whether this response was triggered by reduced oxygen (due to the increased temperature), by increased temperature, or by both of

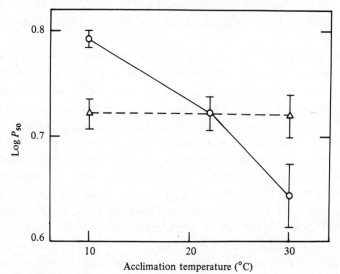

Fig. 2. The effect of temperature on the P_{50} values of blood from fish (*Fundulus heteroclitus*) acclimated to 10, 22 or 30 °C for 30 days. The open circles are for equilibrium measurements made at 22 °C while the open triangles are for blood–oxygen equilibrium measurements made at the temperature of acclimation. Bars indicate $\pm$ S.E.M.

these variables. Therefore, fish were acclimated to various temperatures (10, 22 or 30 °C) but with oxygen concentration maintained constant (about 7 p.p.m.) at each temperature. These fish also showed decreased erythrocyte ATP levels with increased temperature (group II, Table 1). Moreover, fish acclimated to 10 °C but in air-saturated water (12.5 p.p.m.) showed the same erythrocyte ATP levels as those maintained at 10 °C but with 7 p.p.m. oxygen (Table 1). These data clearly demonstrate that the ATP response can be elicited by *increased temperature alone* and is independent of dissociated oxygen in the range 7 to 12.5 p.p.m.

The indirect effect of ATP changes

There is a significant decrease in blood pH with elevated temperature (Reeves, 1977). Moreover, the true pH difference between cold- and warm-acclimated *F. heteroclitus in vivo* is greater than the difference reported here when the pH measurements are made at the temperature of acclimation. A decrease in pH will tend to decrease oxygen–haemoglobin affinity via the Bohr effect. However, a decrease in erythrocyte ATP causes an increase in erythrocyte pH due to the influence of ATP on the Donnan distribution of protons across the membrane. Since both plasma pH and erythrocyte ATP are higher at 10 °C, the decrease in erythrocyte ATP at 30 °C may counter the decrease in erythrocyte pH. The net result may be to 'buffer' or minimise the effect of temperature on cellular pH despite changes in plasma pH. Experiments are under way to test this hypothesis.

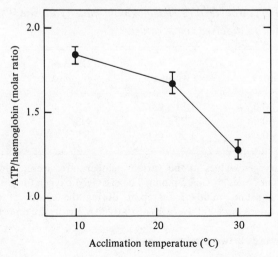

Fig. 3. Molar ratios of ATP/haemoglobin (tetramer) from the blood of *Fundulus heteroclitus* acclimated to 10, 22 or 30 °C for 30 days. Bars indicate ±S.E.M.

Whereas the specific allosteric effect of ATP depends upon the total amount of ATP present with the haemoglobin, the non-specific effects just discussed depend only upon the free, or unbound ATP. Since the binding of ATP to deoxyhaemoglobin will alter the concentration of free ATP, the temperature, oxygen, and pH dependence of the binding equilibrium between haemoglobin and ATP and the state of ionisation (Greaney et al., 1980) must be considered when discussing adaptative mechanisms of the non-specific effects of ATP. In other words, the specific and non-specific effects are intimately linked to one another.

Strategies of adapting to hypoxia

There are numerous strategies by which fish are able to maintain respiratory homeostasis when environmental oxygen is reduced. Perhaps the most common is to seek out a more favourable environment. Species that remain in oxygen-poor environments have (1) immediate, (2) intermediate and (3) long-term responses.

Types of responses

Immediate response. The immediate response to an oxygen-poor environment is an increase in heart rate and ventilation volumes (Prosser, 1973). In addition, fish will often 'gulp' air and/or utilise primarily the water at the air–water interface, which has the greatest oxygen content. These and hormone mechanisms (Powers, unpublished) are usually adequate for short-term hypoxia, but when low oxygen levels are maintained for more than a few hours, additional acclimatory mechanisms are activated.

Table 1. *Comparison of erythrocyte ATP/haemoglobin molar ratios in F. heteroclitus acclimated to different temperatures*[a]

Temperature (°C)	Group I	Group II	Level of significance[b]
10	1.84±0.05	1.81±0.07	NS
22	1.68±0.06	1.72±0.08	NS
30	1.31±0.08	1.20±0.06	NS

[a] In group I, dissolved oxygen values at the various temperatures were: 10 °C, 12.5 p.p.m.; 22 °C, 9.0 p.p.m.; and 30 °C, 7.5 p.p.m. The dissolved oxygen for group II was maintained at a constant value of 7.0 p.p.m. during the course of the acclimation (4 weeks). All values represent averages of 10–12 fish. Errors are reported as ±S.E.M.
[b] NS, no significant differences between groups I and II.

Intermediate responses. Intermediate responses are generally activated after several hours of hypoxic conditions and last many days or until long-term compensation is achieved. These responses include increasing haematocrit by retaining serum in muscle tissues (Cameron, 1970), or releasing stored erythrocytes from the spleen, reducing intra-erythrocyte concentrations of organic phosphates (e.g. Wood & Johansen, 1972; Greaney & Powers, 1978), changing pH, and changing the ionic microenvironment of erythrocytes (Houston & Cyr, 1974; Houston & Mearow, 1979). Such attempts to maintain oxygen delivery to respiring tissues are usually accompanied by large fluctuations in various enzyme activities during metabolic readjustment. During the intermediate response, fish synthesise new erythrocytes so that the total oxygen-carrying capacity of the blood is increased. Eventually, a new steady state between erythrocyte numbers, enzyme levels, organic phosphates and haemoglobin function is achieved.

Response mechanisms. When exposed to low oxygen environments for several days, both mammals (Johansen, 1970) and fish (Krogh & Leitch, 1919) increase the oxygen-carrying capacity of their blood. The mechanisms common to mammals and fish are: increased haematocrit, increased haemoglobin content and increased blood buffering capacity. On the other hand, mammal and fish differ considerably in other aspects of their response to low oxygen. Mammals decrease the affinity of their haemoglobin for oxygen. Hypoxic fish, in contrast, increase oxygen–haemoglobin affinity as characterised by a decrease in the P_{50} of their corresponding oxygen-saturation curves. For example, mammals acclimated to high altitudes, increase levels of 2,3-DPG in their erythrocytes (Lenfant & Johansen, 1968). Since mammals typically live in an oxygen-rich environment, it has been suggested that this response might be best suited to the more commonly encountered forms of low-altitude hypoxia, such as chronic hypoxaemia (Eaton, 1974). The response of water-breathing vertebrates to chronic hypoxaemia is quite different. For example, eels have been shown to decrease erythrocyte ATP and increase oxygen–haemoglobin affinity when acclimated to low environmental oxygen (Wood & Johansen, 1972). In addition, *Fundulus heteroclitus* that have been acclimated to hypoxic conditions reduce the erythrocyte ATP concentrations by as much as 40% and increase the haematocrit, which presumably increases the oxygen-carrying capacity of the blood (Fig. 4). Moreover, there are concomitant increases in serum lactate and a decrease in blood pH (Table 2).

Table 2. *Blood respiratory properties of hypoxic and normoxic F. heroclitus after 13 days of acclimation*

Parameter	Hypoxic	Normoxic	Level of significance
ATP/haemoglobin (tetramer) molar ratio	1.21±0.05 (7)	1.80±0.06 (8)	$P < 0.001$
Blood haematocrit (%)	36±1.0 (7)	23±1.15 (8)	$P < 0.001$
Serum lactate (mM)	6.63±1.22 (6)	3.48±0.90 (15)	$P < 0.001$
pH	7.4±0.1 (5)	7.7±0.5 (10)	$P < 0.001$

The number of experiments is given in brackets below each value.

Fig. 4. *Fundulus heteroclitus* hypoxia studies. (*a*) Time course of blood haematocrit of hypoxic (filled squares) and normoxic (open squares) fish. (*b*) Time course of acclimation of the fish to hypoxic (filled circles) and normoxic (open circles) conditions at 22 °C. Dissolved oxygen values were 0.2–2.0 p.p.m. for hypoxic fish and 8.5–9.0 p.p.m. for control fish. Erythrocyte ATP concentrations for this experiment and all others described in this paper were determined using the firefly luciferase assay. All points represent an average of 6–7 fish. Bars indicate ± S.E.M.

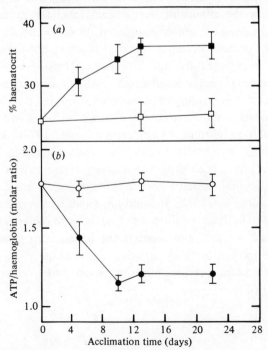

Cellular regulation of ATP concentration

One of the obvious questions would be: 'Is this response to hypoxia regulated at the organism level, erythrocyte level, or both?' If the control were directed at the erythrocyte level, then fish erythrocytes should decrease ATP *in vitro* under anoxic conditions. Consistent with our observation *in vivo* (Fig. 4), we found that *F. heteroclitus* erythrocytes significantly lowered their ATP levels *in vitro* (Fig. 5). Since the nucleated erythrocytes of fish possess mitochondria, we reasoned that this response may be mediated by way of a decrease in the rate of oxidative phosphorylation. This hypothesis was supported when aerated cells were incubated in the presence of low concentration of cyanide (Fig. 6). This inhibitor of aerobic respiration reduced intracellular ATP to levels similar to those found in the anoxic cells. Moreover, oxygen consumption by *Fundulus* erythroyctes was eliminated by the presence of cyanide (Fig. 7). Thus it seems that one reason that fish might have retained a functional oxygen-consuming electron transport system in the mitochondria of their erythrocytes would be to control the concentration of ATP, an allosteric effector of haemoglobin.

Glycolytic inhibition experiments were consistent with the cyanide data (Fig. 6). With iodoacetate, an inhibitor of glyceraldehyde-3-phosphate dehydrogenase (GAPDH), reducing equivalents from glycolysis are not available for oxidation by the mitochondrial electron transport chain and less pyruvate is oxidised in the tricarboxylic acid (TCA) cycle. Consequently,

Fig. 5. ATP/haemoglobin molar ratio of *F. heteroclitus* erythrocytes incubated under anaeroboic (filled circles) and aerobic (open circles) conditions. Cells were washed three times in phosphate-buffered saline (pH 7) and then maintained in a medium containing physiological concentrations of inorganic salts and glucose buffered at pH 7.5, as described elsewhere (Greaney & Powers, 1978). Values are from duplicate determinations which do not vary by more than 6%.

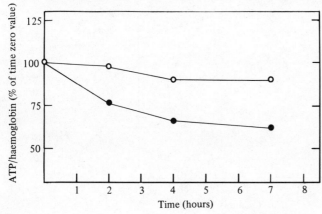

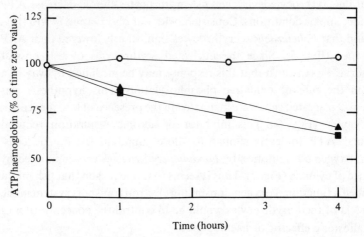

Fig. 6. Time course of changes in the ATP/haemoglobin ratio in cells maintained aerobically at pH 7.5 in Bull media with no inhibitor (open circles), 10^{-4} M sodium cyanide (filled squares) or 5×10^{-4} M iodoacetate (filled triangles).

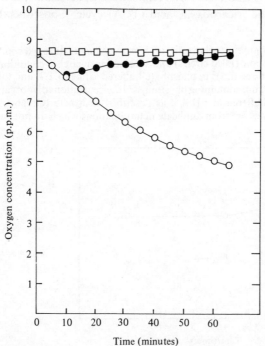

Fig. 7. Oxygen consumption of *Fundulus heteroclitus* erythrocytes (10% haematocrit) in the presence (open circles) or absence (filled circles) of 10 mM cyanide. A human erythrocyte control (open squares) is included for comparison.

erythrocytes exposed to this inhibitor lower their ATP to levels nearly equal to those of cyanide-poisoned cells. In the presence of fluoride, which inhibits enolase, the erythrocytes showed concentrations of ATP only slightly lower than control levels since reducing equivalents are formed during the earlier GAPDH reaction which can then fuel mitochondrial oxidative phosphorylation.

Assuming that iodoacetate and fluoride are acting as specific inhibitors in the erythrocytes, these data indicate that the small amount of reducing equivalents from glycolysis would be nearly identical to that resulting from a fully functioning TCA cycle. Recent studies have clearly demonstrated that erythrocytes of *Fundulus heteroclitus* have active TCA cycle enzymes within their mitochondria and these erythrocytes consume significant amounts of oxygen. As mentioned above, cyanide eliminates the ability of these erythrocytes to consume oxygen (Fig. 7).

Metabolic partitioning in fish erythrocytes

In order to understand further how erythrocyte ATP levels are controlled, and to estimate the relative participation of the TCA cycle, pentose shunt and glycolytic pathways, we investigated the metabolism of radiolabelled substrates in *Fundulus* erythrocytes *in vitro*. The rate of $^{14}CO_2$ evolution was monitored either by the continuous-recording method of Larrabee (1977) or by a manual modification of Larrabee's automated procedure. Lactate was measured after chromatographic separation.

By using glucose or lactate radiolabelled in appropriate positions (C-1, C-2, C-3, C-6, etc.) one can determine the partitioning of carbon into the various pathways. Those experiments indicate that fish (*F. heteroclitus*) erythrocytes partition approximately 76% of the carbon from glucose into glycolysis and 24% into the pentose shunt. Of the 76% partitioned into glycolysis, 71% ended up as lactate and 5% was directed into the TCA cycle and oxidised to $^{14}CO_2$. When [^{14}C]lactate was labelled in various positions, it was readily metabolised to $^{14}CO_2$. This was true even when various steps in glycolysis were blocked with metabolic poisons. However, cyanide significantly reduced the metabolism of [^{14}C]lactate, presumably at the first decarboxylation step. These data implicate a small but significant participation of the TCA cycle in fish erythrocyte metabolism.

Since oxygen concentration directly affects oxidative phosphorylation and the yield of ATP via the TCA cycle and electron transport, it is reasonable to expect that some of the changes in erythrocyte ATP in response to changes in environmental oxygen and/or temperature may be accounted for by the small (5%) but significant mitochondrial activity.

On the other hand, when isolated erythrocytes are exposed to 10%

air-saturated medium no decrease in ATP from the control value is apparent even after 8 hours. This suggests that the erythrocyte metabolism is only anaerobic when most of the oxygen is removed. It is reassuring that the cytochrome oxidase system of mammalian aerobic respiration is known to have a very low Michaelis constant (K_m) for oxygen and appears to be saturated with oxygen *in vivo* except under conditions of extreme oxygen lack.

In vivo erythrocytes are only anoxic in the venous blood of hypoxic fish. Thus, the fraction of time exposed to anoxic conditions is much less *in vivo* than *in vitro*. The decrease in erythrocyte ATP levels *in vivo* is not evident until 5 to 10 days after exposure to 1.5 p.p.m. oxygen. It seems reasonable that this response is mediated at the erythrocyte level via decreased oxidative phosphorylation in response to lower environmental oxygen. Since *in vivo* only the venous blood is anoxic, the decrease in ATP is not observed for several days at the oxygen concentration used in the whole-animal acclimation experiments. *In vitro* the response is accelerated and can be observed in a matter of hours.

Significance of haematocrit alterations

The increasing haematocrit observed in hypoxic *F. heteroclitus* (Fig. 4a) is similar to measurements made on hypoxic eels (Wood & Johansen, 1972) and mammals (Prosser, 1973). The adaptive nature of the response is interpreted to be an increased oxygen-carrying capacity of the blood under conditions of lowered environmental oxygen. Since qualitative and quantitative correlations exist between the appearance of new erythrocytes and the decrease in the ATP/haemoglobin ratio, it is important to consider their possible significance. Erythrocytes that are being released into the circulatory system might conceivably have lower intracellular ATP concentrations than older cells. Therefore, a sampling of the entire pool of erythrocytes will give an apparently lower average ATP/haemoglobin ratio per erythrocyte. This hypothesis, however, is not supported by the observation of Wood, Johansen & Weber (1975) that plaice, *Pleuronectes platessa*, raise their erythrocyte ATP values when they are removed from their natural, semi-anaerobic environment and exposed to aerated water. Additional evidence against this possibility is that younger human erythrocytes have higher, not lower, concentrations of ATP.

Of the two hypotheses discussed, we favour the former and suggest that, during hypoxic acclimation, fish increase oxygen affinity by lowering their erythrocyte ATP levels via a direct cellular response to lower environmental oxygen. This results from a decreased production of ATP via oxidative phosphorylation. The response is not apparent for several days at the oxygen concentrations used in our acclimation studies, presumably because erythrocytes are only anoxic in the venous blood. The increased haematocrit is then simply a mechanism for increasing the total oxygen-carrying capacity of the blood so that homeostasis may be maintained during long-term acclimatisation to hypoxic environments.

Summary

Animals live in environments where physical, chemical and biological parameters are continually changing. If homeostasis is to be maintained, these animals must adapt. The respiratory complex, and specifically haemoglobin and its ligand interactions, is perhaps the best system to study this type of adaptation because it exists at the 'organism–environment interface'. Fish are particularly useful in such studies because they respond to the environmental variables mentioned above.

The mechanisms by which respiratory homeostasis is maintained involve molecular, cellular and physiological levels of organisation. Molecular adaptation includes adjustments in the concentrations of ligands that are thermodynamically linked to oxygen–haemoglobin affinity. We have shown that fish erythrocytes are capable of adjusting the concentrations of their intracellular organic phosphates, protons and other constituents when environmental temperature and/or oxygen changes. Fish can also increase the number of circulating erythrocytes thereby increasing or maintaining the oxygen-carrying capacity of the blood in an oxygen-poor environment. These and other changes (e.g. hormonal, ionic) facilitate adaptation of the oxygen transport system which is vital to survival.

This work was supported by NSF grant DEB 79-12216. Additional support was provided by an NIH Biomedical Grant to JHU. A portion of this work was done on two cruises on the RV *Alpha Helix* supported by NSF grants PCM 75-06751 and PCM 76-06451. This is contribution no. 1184 from the Department of Biology, and the McCollum-Pratt Institute, the Johns Hopkins University, Baltimore, Maryland 21218.

References

Benesch, R. & Benesch, R. E. (1967). Effect of organic phosphates from the human erythrocytes on the allosteric properties of hemoglobin. *Biochemical and Biophysical Research Communications*, **26**, 162–70.

Bonaventura, J., Bonaventura, C. & Sullivan, B. (1975). Hemoglobins and hemocyanins: comparative aspects of structure and function. *Journal of Experimental Zoology*, **194**, 155–74.

Brunori, M. (1975). Molecular adaptation to physiological requirements: the hemoglobin system of trout. *Current Topics in Cellular Regulation*, **9**, 1–39.

Cameron, J. N. (1970). The influence of environmental variables on the haematology of the pinfish (*Lagodon rhomboides*) and striped mullet (*Mugil cephalus*). *Comparative Biochemistry and Physiology*, **32**, 175–92.

Eaton, J. W. (1974). Oxygen affinity and environmental adaptation. *Annals of the New York Academy of Sciences*, **241**, 491–7.

Fry, F. E. J. & Hart, J. S. (1948). The relation of temperature to oxygen consumption in the goldfish. *Biological Bulletin* (*Woods Hole, Mass.*), **94**, 66–77.

Geohegan, W. D. & Poluhowich, J. J. (1974). The major erythrocyte organic phosphates of the American eel *Anguilla rostrata*. *Comparative Biochemistry and Physiology*, **49B**, 281–90.

Gillen, R. G. & Riggs, A. (1971). The hemoglobins in a fresh-water teleost *Cichlasoma cyanoguttatum*: the effects of phosphorylated organic compounds upon oxygen equilibria. *Comparative Biochemistry and Physiology*, **38B**, 585–95.

Greaney, G. S., Hobish, M. K. & Powers, D. A. (1980). The effects of temperature and pH on the binding of ATP to carp (*Cyprinus carpio*) deoxyhemoglobin (HbI). *Journal of Biological Chemistry*, **255**, 445–53.

Greaney, G. S. & Powers, D. A. (1978). Allosteric modifiers of fish hemoglobin: *in vitro* and *in vivo* studies of the effect of ambient oxygen and pH on erythrocyte ATP concentrations. *Journal of Experimental Zoology*, **203**, 339–50.

Greaney, G. S. & Powers, D. A. (1979). Cellular regulation of an allosteric modifier of fish hemoglobin. *Nature* (*London*), **270**, 73–4.

Haugaard, N. (1974). The effect of high and low oxygen tension on metabolism. In *Molecular Oxygen in Biology*, ed. G. Hajarshi, pp. 163–80. New York: American Elsevier.

Houston, A. H. (1980). Components of the hematological response of fishes to environmental temperature change: a review. In *Environmental Physiology of Fishes*, ed. M. A. Al, pp. 241–98. New York: Palum Publishing Corp.

Houston, A. H. & Cyr, D. (1974). Thermoacclimatory variation in the hemoglobin system of goldfish (*Carassius auratus*) and rainbow trout (*Salmo gairdneri*). *Journal of Experimental Biology*, **61**, 455–61.

Houston, A. H. & Mearow, K. A. (1979). Temperature related changes in the erythrocyte carbonic anhydrase (acetazolamide-sensitive esterase) activity of goldfish, *Carassius auratus*. *Journal of Experimental Biology*, **78**, 255–64.

Johansen, K. (1970). Airbreathing in fishes. In *Fish Physiology*, vol. 4, ed. W. S. Hoar & D. J. Randall, pp. 361–411. New York & London: Academic Press.

Johansen, K. & Lenfant, C. (1972). A comparative approach to the adaptability of [O_2]-Hb affinity. In *Oxygen Affinity of Hemoglobin and Red Cell Acid–Base Status*, ed. P. Astrup & M. Rorth. Copenhagen: Munksgaard.

Krogh, A. & Leitch, I. (1919). The respiratory function of blood in fishes. *Journal of Physiology* (*London*), **52**, 288.

Larrabee, M. G. (1977). Continuous measurement of labeled carbon dioxide output from small tissue samples, using a low-background system. *Analytical Biochemistry*, **79**, 357–9.

Lenfant, C. & Johansen, K. (1968). Respiration in the African lungfish, *Protopterus aethiopicus*. I. Respiratory properties of blood and normal patterns of breathing and gas exchange. *Journal of Experimental Biology*, **49**, 437–52.

Mied, P. & Powers, D. A. (1977). Hemoglobins of the killifish *Fundulus heteroclitus*: separation, characterization and a model for the subunit composition. *Journal of Biological Chemistry*, **253**, 3521–8.

Powers, D. A. (1972). Hemoglobin adaptation for fast and slow water habitats in sympatric catostomid fishes. *Science*, **177**, 360–2.

Powers, D. A. (1974). Structure-functions and molecular ecology of fish hemoglobins. *Annals of the New York Academy of Sciences*, **241**, 472–90.

Powers, D. A. (1980). Molecular ecology of teleost fish hemoglobins: strategies for adapting to changing environments. *American Zoologist*, **20**, 139–62.

Powers, D. A. & Edmundson, A. B. (1972). Multiple hemoglobins of catostomid fish. I. Isolation and characterization of the isohemoglobins from *Catostomus clarkii*. *Journal of Biological Chemistry*, **247**, 6686–93.

Powers, D. A., Fyhn, H. J., Fyhn, U. F. H., Martin, J. P., Garlick, R. L. & Wood, S. C. (1979a). A comparative study of the oxygen equilibria of blood for 40 genera of Amazonian fishes. *Comparative Biochemistry and Physiology*, **62A**, 67–85.

Powers, D. A., Martin, J. P., Garlick, R. L. & Fyhn, H. J. (1979b). The effect of temperature on the oxygen equilibria of fish hemoglobins in relation to environmental thermal variability. *Comparative Biochemistry and Physiology*, **62A**, 87–94.

Powers, D. A. & Powers, D. (1975). Predicting gene frequencies in a natural population: a testable hypothesis. In *The Isozymes IV, Genetics and Evolution*, vol. 4, ed. C. Markert, pp. 63–84. New York & London: Academic Press.

Prosser, C. L. (1973). *Comparative Animal Physiology*, 3rd edn. Philadelphia: W. B. Saunders.

Reeves, R. B. (1977). The interaction of body temperature and acid–base balance in ectothermic vertebrates. *Annual Review of Physiology*, **39**, 559–86.

Riggs, A. (1970). Properties of fish hemoglobins. In *Fish Physiology*, vol. 4, ed. W. S. Hoar & D. J. Randall, pp. 209–52. New York & London: Academic Press.

Riggs, A. (1971). Mechanism of the enhancement of the Bohr effect in mammalian hemoglobins by diphosphoglycerate. *Proceedings of the National Academy of Sciences, USA*, **68**, 2062–5.

Root, R. W. (1931). The respiratory function of the blood of marine fishes. *Biological Bulletin (Woods Hole, Mass.)*, **61**, 427–56.

Weber, R. E., Sullivan, B., Bonaventura, J. & Bonaventura, C. (1976). The hemoglobin system of the primitive fish *Amia calva*: isolation and functional characterization of the individual hemoglobin components. *Biochimica et Biophysica Acta*, **434**, 18–31.

Wood, S. C. & Johansen, K. (1972). Adaptation to hypoxia by increased HbO_2 affinity and decreased red cell ATP concentration. *Nature (London)*, **237**, 278–9.

Wood, S. C., Johansen, K. & Weber, R. E. (1975). Effects of ambient p_{O_2} on O_2–Hb affinity and red cell ATP concentration in a benthic fish, *Pleuronectes platessa*. *Respiratory Physiology*, **25**, 259–67.

Wyman, J. (1964). Linked functions and reciprocal effects in hemoglobin: a second look. In *Advances in Protein Chemistry*, vol. 19, ed. C. B. Anfinsen, M. L. Anson, J. T. Edsall & F. M. Richards, pp. 223–86. New York & London: Academic Press.

**M. W. SMITH, P. SHETERLINE
and A. R. COSSINS**

Conclusions: a cellular perspective on environmental physiology

It is evident from the foregoing discussions that the cellular mechanisms of acclimatisation occupy the efforts of a large number of scientists from many different disciplines. Much of the current enthusiasm comes from the somewhat surprising discovery that such diverse work should now be producing a consensus view about the nature of the strategies used by cells when adjusting to changes in their immediate environment. It is no longer fanciful, for instance, to stress the common features of cellular adaptation to temperature and drugs, or to suggest that there are now many findings which describe similar fundamental mechanisms employed by cells in response to changes in oxygen tension, temperature and salinity. All of these subjects have been covered in this volume. The question that might now be asked is whether it is possible to conceive of what is already known about cellular responses as a universal and integrated set of adaptive mechanisms.

A significant change in emphasis may be perceived in what was once a rather fragmentary field for the environmental physiologist. This reflects an understanding that it may no longer be appropriate to consider such adaptive responses of cells *only* in the context of the environmental stimuli to which they respond. Instead, cellular acclimatisation may be a manifestation of widespread mechanisms by which cells establish and maintain an appropriate steady state, even in animals whose internal environment is maintained constant and whose cells may never experience particular environmental insults. For example, the constant core temperature of mammals does not appear to exclude the possession of a cellular mechanism to control membrane fluidity.

One of the experimental difficulties which arises in trying to understand which of the observed cellular responses are adaptive, comes in the first instance from the lack of information on the temporal relationships of those changes which lead to adaptation. One might imagine that the first changes seen to take place in a cell will occur as a direct consequence of the environmental insult, rather than as a cellular response for survival. These initial changes presumably act as initiators of organised responses by the cell which then take time to complete. Whether the final response takes place as a series of timed events or whether several organised changes take place with the same time-course remains unknown. There are, also, second-order events which take place within cells following these initial responses which ensure survival. These have been referred to as the 'settling down effects' in relation to homeoviscous adaptation, where the composition of phospholipid head groups in membranes shows delayed changes which, nevertheless, precede full resumption of normal cell function. The implication here is that delayed effects apply the finishing touches to a survival process organised within a shorter period of time. Another way of describing this type of change, however, is to suggest that the earlier events represent a crisis response to environmental stress – an organised response which acts as an umbrella under which further adaptive changes can be initiated. This would be more in keeping with the early regulation of cell potassium seen to take place in the intestines of fish exposed to different temperatures. In this case it is the leak process which is regulated. The later alteration in sodium/potassium pump activity then makes this earlier reduction in passive permeability unnecessary.

At what point after the environmental change should we become interested in those organised events which lead to cellular adaptation? The answer might be the shortest time needed for an enzyme to re-establish a new (and presumably advantageous) steady state, or perhaps for transcriptional and translational changes to create an appropriate set of proteins. Knowing this, however, still leaves a question as to where to look for subsequent alterations in the cellular function. Some approaches to the identification of potential regulatory sites which might elicit particular adaptive responses have been referred to in this volume and it may be worthwhile to indulge in more of this intelligent guesswork. Apart from intermediary metabolism there is another obvious site for adaptive control and that is in the barrier properties of cellular membranes. Control of the flux of nutrients, coupled with the ability to maintain a high relative concentration of intracellular potassium and a low concentration of intracellular free calcium ions, through the control of active pumps and leakage pathways, is likely to be a prerequisite for solving problems of lesser importance. This, of course, relies upon the continued provision of sufficient ATP to drive cation pumps. The availability or otherwise of ATP immediately following a change in the environment has not, to our knowledge, been considered worthy of study in the past. Such a study could provide important information in relation to the role of ATP-dependent functions in cellular adaptation, particularly if it also involved a study of systems which failed to survive changes in their immediate environment.

Further guesswork might be applied to identify the level of control of cellular acclimatisation. Adaptation which maintains cellular homeostasis, for example, in response to changes in body temperature, are probably mediated at the cellular level. By contrast, cellular responses which promote whole-animal homeostasis might result from the adaptation of specialised cells either directly or through the action of hormones or nerves. Good illustrations of this latter phenomenon seem to be the seasonal induction of antifreeze proteins in winter flounder *before* the occurrence of freezing temperatures, and secondly the influence of cortisol and prolactin upon the osmoregulatory and structural characteristics of gill epithelium of fish. At present there is very little unequivocal information concerning the control mechanisms for any of the physiological adaptive responses described in this volume.

It could be argued that this somewhat anthropomorphic way of describing how cells might operate at times of environmental stress is positively misleading and that what is needed in future is a deeper appreciation of one or other of the many different aspects of cellular adaptation studied at the present time. Our only defence against this argument is to remind you that the title of this volume is *Cellular Acclimatisation to Environmental Change*. We may find ourselves working on systems which vary in complexity from pure enzymes to intact organisms, but this should not, we think, provide an excuse for ignoring the challenge of consciously attempting to describe all our work in terms of how the whole cell adapts to its immediate environment.

INDEX

acclimation 1
  adaptive changes in carp 23
  lipid role 33
  membrane restructuring and adaptive change 16
  phospholipid reorganisation 42
acclimatisation 1
  enzyme changes 103–18
  in vitro enzyme activities 81–99
  levels of control 246, 247
acetylcholinesterase mobility and temperature in trout 134, 135
actomyosin properties and temperature 138
adaptive responses
  categories 1, 2
  definition 179
  membrane relationships 47, 48
  salt 199, 200
adenosine, inhibitory receptor 171
adenylate cyclase
  hamster liver 21, 22
  morphine tolerance in neuroblastoma × glioma hybrid 170, 171
aflatoxin B_1 188, 189
alanine
  in antifreeze peptides 217, 218, 224
  degradation and hypo-osmotic conditions 70
alcohols *see also* ethanol
  chain length and *E. coli* growth 148
  membrane fluidisation 148, 149
alfalfa, chloroplast fatty acid unsaturation 35
alloforms 122
amino acid, role in volume-regulatory decrease 69–71
antifreeze peptides and glycopeptides
  biosynthesis regulation 220, 221
  DNA cloning 221–5
  DNA nucleotide sequence 223
  gene numbers and properties 224, 225
  ice binding properties 224
  mechanism of action 218, 219

  mRNA isolation 220, 221
  seasonal variations 219, 220, 247
  species specificity 217
  structures 218
  synthesis regulation 217–26
Arrhenius discontinuities and phase transitions 21
arylcarbon hydroxylase, microsomal and genetic variance 188
ATP
  erythrocyte concentration in fish: and hypoxia 237–41; and temperature 232, 233
  fish haemoglobin binding 230, 231
  luciferase assay 236
ATPase, Mg^{2+}
  negative temperature dependence 19
  temperature dependence in fish 136, 137
ATPase, myofibrillar properties
  calcium and temperature 138
  temperature 135
ATPase, $Na^+ + K^+$
  cholesterol role 17, 18
  inactivation and temperature 19
  seawater adaptation in fish 202
  specific activity and acclimation 19
  thermal stability in goldfish 20
  transport function and temperature 22, 23

Bacillus megaterium desaturase and temperature 39
benzo(a)pyrene metabolism and monooxygenase induction 187, 191
biotransformation pathways 179
biphenyl metabolism and monooxygenase induction 187, 191
birds, fat flight fuel 98
bladder, urinary in fish: salt adaptation control 199, 208
body temperature
  altered and protein isoforms 121–39
  pH effects in fish 125–7
Bohr effect 228

INDEX

Boyle–Van't Hoff law 61
bumetanide 67
   and co-transport systems 71, 73
   steady-state chloride influx 75

caffeine elimination in rats and pretreatment 187, 190
Carassius auratus (goldfish)
   ATPase, myofibrillar and temperature 136, 137
   n-butanol tolerance 156
   cytochrome content and temperature 112
   fatty acid unsaturation 35
   mitochondria in muscle and temperature 116, 117
   phospholipid head groups and temperature 44, 45
   seawater tolerance and adaptation 208
   time course of membrane fluidity changes 10, 11
Carassius carassius (crucian carp), mitochondria and temperature 115
carcinogenesis, chemical and monooxygenase inducers 188, 189, 194
carnitine palmitoyl transferase 99
carp, Arrhenius discontinuities in liver enzymes 21
cells *see also* chloride cells
   Chinese hamster ovary 150
   hypotonic media, volume regulation 62, 63
   ion permeability 61
   mammalian *in vitro* 149–51
   neuroblastoma × glial, ODT 170, 171
   opiate dependence 163
   water permeability 60, 61
cell volume *see also* volume regulation
   and calcium 68
   -dependent anion transport 68, 69
   membrane selective permeability ratio 66
   osmolarity 63, 64
   osmotic behaviour 60–2
   physical principles 55
   potassium influx 74
chloride, cell content and osmolarity 67
chloride cells 200–5
   hormonal effects 205
   salt secretion model 204
   and seawater adaptation in fish 202, 203
   subcellular changes and salinity adaptation 202, 203
cholesterol and membrane ethanol tolerance 154, 156
citrate synthase, maximal activity and species 89–93
cold acclimation
   capillary supply 132
   enzyme changes in carp 131, 132
cold stress, metabolic strategies 128

cortisol and salinity change in fish 209
cytochromes
   content in fish and temperature 104, 111–14
   extraction and analysis 113
cytochrome(s) P_{450}
   definition 180
   inducers 182
   inducers and metabolic profile 187, 191
   induction 180–5
   induction of hepatic 179–95
   microsome content and inducer 182
   multiple forms 180
   prenatal induction 191, 194
   species differences in induction 182, 185
   steroid hormone metabolism 190, 191, 194

dependence
   receptor theory for opiates 168, 169
   and tolerance 161, 162
   uncertainty principle 173
desaturases 36, 37
   regulation and membrane fluidity 38, 39
   synthesis and temperature 38
diphenylhexatriene, polarisation and temperature in *Tetrahymena* cells 47, 49
DNA cloning for antifreeze peptides 221–5
Donnan equilibrium 57, 58
drugs
   cytochrome P_{450} induction 180
   depressant 155
   and membrane changes 145–57
Dunaliella salina, lipid polar head groups and stress 44–6

eels, seawater and drinking reflex 209
Ehrlich ascites tumour cells 56
   anion flux 68
   chilling and electrolyte uptake 59
   cotransport systems 74
   ion and water permeability 61
   volume-regulatory decrease 62–71
   volume-regulatory increase 70–6
   volume-regulatory responses 62, 70
electron spin resonance spectroscopy 4
enzyme activities
   allosteric activation 105, 127
   elevation of concentration 105, 106, 128
   enhanced in cold acclimation 87
   in vitro and acclimatisation 81–99
   maximum as flux indicators 82–5
   temperature responses 127–9
enzymes
   changes and environmental stimuli 104
   complexes 106
   concentration alteration 107, 108
   concentration measurement 108

enzymes (*cont.*)
  flux-generating reaction 83–5
  interconversion 107
  near-equilibrium reaction 82, 83
  non-equilibrium reaction 82, 83
  quantitative changes 103–18
  solubility 106
  subunit composition 134
  temperature acclimation in fish 111
erythrocyte
  ATP:haemoglobin ratio in hypoxia 240
  metabolic partitioning 239, 240
  oxygen consumption and cyanide 237–9
erythrocyte adaptation
  environmental changes 227–41
  in hypoxia 236, 237
Escherichia coli
  DNA cloning 222, 223
  fatty acid synthesis regulation 39, 40
  fatty acid unsaturation 34, 35
  homeoviscous adaptation to drugs 148
  homeoviscous efficacy 9
  membrane phase separation 12, 13
  membrane transport and temperature 22, 23
ethanol
  administration and effect 154
  mammalian cell culture effects 150, 151
  membrane lipid accumulation 152, 153
  sensitivity and body weight 157
  tolerance: and cholesterol 156; and membrane lipids 147, 152–4; and temperature in mice 155
eukaryotes
  homeoviscous efficacy 8, 9
  membrane types 24

fatty acids
  branched chain, proportions 40, 41
  chain length changes 41
  desaturation: electron transport 36, 37; pathways in animals and plants 36, 37; in prokaryotes 39
  position in membrane lipids 41–3
  synthesis and temperature 38
  unsaturation and stress 34–6
feedback control, open and closed 25
final cholinergic motoneurone 163
  axon 169
  functional divisions 167
  opiate dependence and associated tolerance 166
  receptor site 166–9
  soma 169, 170
  terminal 169
fish *see also individual fish species*
  cold acclimation, protein and capillary effects 131, 132
  enzyme changes and temperature 111
  fuels 98
  lipid storage 98
  muscle and swimming 97, 98
  oxygen requirements and temperature 231, 232
  salinity adaptation 197–200
  sea- and freshwater, osmoregulation 198, 199
  temperature adaptation 121
fluid 4
flux-generating reaction 83–5
freshwater
  adaptation changes in fish 198, 199
  chloride cells in fish 202, 203
frusemide 67
  co-transport systems 71, 73
fuels
  aerobic comparison 96–9
  fish 97, 98
  insects 97
Fundulus heteroclitus
  blood–oxygen affinity 232
  hypoxia and erythrocyte ATP 235–7

gas chromatography–mass spectroscopy, phospholipid analysis 42
general anaesthetics, membrane effects 155
Gibbs–Donnan distribution 55
  electroneutrality 58
  role in cell volume 57, 58
gills
  absorption and excretion 201
  epithelium role 200, 201
  fresh- and seawater adaptations in fish 198, 205
  salt transport and hormones 205
  structure and function 200, 201
glycopeptides *see* antifreeze peptides and glycopeptides
guinea-pig ileum
  cell size of ODT 165, 166
  opiate dependence characteristics 165, 166
  opiate tolerance and dependence 163, 164
  opiate withdrawal effect 164

haematocrit in hypoxia 235, 236, 240
haemoglobin
  to ATP ratio: and cyanide 237, 238; and temperature 232–4
  functions 227
  ligand binding properties 228, 229
  multiple in fish 227, 228
  organic phosphate binding 230, 231
  oxygen affinity: and hypoxia 235; regulation 229, 230
heart
  double circulation 94
  muscle activity 94

INDEX

hexobarbital, multiple doses and elimination in dogs 185–9
hibernation 127–30
  enzyme changes 129
  isoenzyme composition 129
  phase behaviour adjustment 21
  pyruvate kinase in bats 129, 130
  temperature drop 127
homeoviscous adaptation 3, 6–11
  adaptive significance 12, 14–23
  drug effects on membranes 145
  membrane lipid composition 7, 8, 34–47
  liver mitochondrial membranes 6–8
  mechanisms 33–50
  predicted and actual alcohol response 153
  settling-down effects 246
  and temperature 3–27, 146, 147
  time course 10, 11
homeoviscous drug adaptation hypothesis, membrane changes 146
homeoviscous efficacy
  and acclimation temperature 8
  prokaryote and eukaryote comparison 8, 9
homeoviscous response
  adaptive value 26, 27
  enzyme synthesis 38
  fatty acid desaturation 36, 37
  oxygen supply 37, 38
β-hydroxydecanoyl thioester dehydrase 39
hyperthermia
  and drug tolerance 155–7
  ethanol cross-tolerance 156
hypothermia and drug tolerance 155–7
hypoxia
  adaptation in fish 234–41
  blood oxygen-carrying capacity 235
  blood properties 236
  erythrocyte ATP levels 236, 237
  haematocrit 235, 236
  responses in fish 234, 235

insects
  fat 97
  flight and tricarboxylic acid cycle 97
  haemolymph fuels 97
intestine, fish
  chloride-transporting epithelium 206, 207
  salinity adaptation role 199, 205–8
  salt transport model 206, 207
isocitrate dehydrogenase, maximal activity and species 89–93

lactic dehydrogenase
  characteristics and thermal environment 103
  evolutionary relationship of genes 123, 124
  heart type 123
  Michaelis constant and temperature 126
  temperature acclimation response 123
Lepomis cyanellus (green sunfish)
  cytochrome c and temperature acclimation 112–14
  enzyme changes and temperature 131
  membrane fluidity and temperature 6, 7
  phosphoglucomutase isoenzymes 124, 125
lindane toxicity and cytochrome induction 188, 193
lipid
  class proportions 44, 45
  polar head groups and stress 44, 45
  structural alterations 33
lipid biosynthesis control 26
liposome
  permeability and phospholipids 14, 15, 27
  temperature-acclimated trout 15

mammals, muscle fuels 99
membranes
  biochemical responses to alcohol 152–5
  boundary layer 16
  changes and chronic drug exposure 145–57
  conductive permeability 61
  delipidation 17
  dynamic structure 4, 5
  fluid mosaic model 16
  homeoviscous adaptation 3–27
  phase separation 5
  phase structure adaptations 11, 12
  phase transitions 21, 22
  temperature and permeability 14–16
  thermal stability 20, 21
  transport processes and temperature 22, 23
  zones, structural 5
membrane-bound enzyme
  activity and temperature 16, 17
  and membrane fluidity 17–19
  specific activity and thermal acclimation 19
membrane fluidity
  and alcohols 148–51
  biosynthesis control 26
  desaturase regulation 38, 39
  drug effects 152
  ethanol adaptation 152, 153
  feedback control 25
  maintenance and temperature change rate 47, 48
  optimal 24, 33
  phosphoglycerides 24
  protein dynamics 19
  temperature 3–20

membrane lipids
  fatty acid positional changes 41–3
  polar head group size 46
  stress-induced alterations 33, 34
  and temperature 34
menstrual irregularities and PCB consumption 190, 191
metabolic partitioning in fish 239, 240
metabolic pathway
  and chemical toxicity 187
  definition 84
metabolism
  aerobic, index 86, 87
  fuels 81, 82
methoxyflurane and *Tetrahymena* lipids 149
3-methylcholanthrene
  hepatic necrosis and pretreatment 185, 192
  lindane toxicity 188, 193
  monooxygenase induction and species 185
Micrococcus cryophilus, fatty acid chain length and temperature 41
microsomes, hepatic protein profile and PCB 182, 184
mitochondria
  changes and temperature response 114
  diffusion equation 114, 115
  proportion of cell volume and temperature 115
  red and white muscle and temperature 116, 117
Morone saxatilis (striped bass): temperature acclimation and oxygen utilisation 110
morphine, homeoviscous adaptation effects 155
muscle
  classification and enzyme V_{max} 95
  comparative metabolism 87
  insect flight, power output 94
  TCA cycle and energy provision 88–96
  white anaerobic 96
myosin
  factors affecting 122, 123
  functionally distinct forms 136, 137
  polymorphic forms 122, 123
Myotis lucifugus (little brown bats): pyruvate kinase and temperature 130, 131

naloxone, vas deferens excitability 172
near-equilibrium reactions 82–5
  enzyme activities 85
non-equilibrium reactions 82–5
  and fuel utilisation rate 85, 86
nuclear magnetic resonance spectroscopy 4

ODT *see* opiate dependence and associated tolerance

oesophagus, teleost and salt adaptation 199, 205
opiate dependence and associated tolerance (ODT) 162–73
  guinea-pig ileum 163–5; characteristics 165, 166
  in vitro in neurones 163
  mouse vas deferens 172, 173
  neuroblastoma × glial cells 170, 171
  receptor-mediated function in neurones 162
  receptor theory 168
  tolerance 162–73
opiates
  adaptation 161, 162
  antagonists 163
  binding sites and tolerance 167
  disinhibition 162
  endogenous 161
  receptor types 162
organelles, changes during acclimatisation 114–18
osmolarity
  cell volume and potassium and chloride changes 72
  changes and cell volume 64
  potassium permeability 65
  seawater 197
osmoregulation
  organs 197
  teleost 197–200
osmotic behaviour, perfect 61, 62
ouabain 56
  and cell volume 60
oxoglutarate dehydrogenase
  activities in muscle 88
  activity, species comparison 88–96
  aerobic metabolism index 86, 87
  birds 92, 93, 98, 99
  and double circulation 94
  fish 91, 92, 97, 98
  insects 89–91, 97
  mammals 93, 99
  TCA cycle flux 86–8
  V_{max} and muscle classification 95, 96
oxygen
  desaturase activity 37, 38
  diffusion coefficient and temperature 115
  -poor environments and fish 234
  utilisation and thermal acclimation in fish 110

paracetamol and mortality in pretreated mice 188, 192
PCB *see* polychlorinated biphenyls
peptides *see* antifreeze peptides and glycopeptides
pH
  blood regulation 125
  and fish body temperature 125–7

phosphates, organic: haemoglobin binding 230, 231
phosphatidylethanolamine, fatty acids and chilling 42, 43
phosphoglucomutase, muscle isoenzymes 124, 125
phospholipids 22
  analysis 42
  expanded at low temperature 7
  unsaturation: and liposome permeability 14, 24; and temperature 36
piretanide 207
polybrominated biphenyls, monooxygenase induction 190, 191
polychlorinated biphenyls
  cytochrome-inducing agents 181
  hepatic monooxygenase induction 181, 183
  monooxygenase effects in trout 184, 186
  species response variation 182, 184, 185
  structures 181
  suckling transfer and induction 194
potassium
  cell permeability and osmolarity 65
  volume-dependent movement 66–8
prokaryotes
  fatty acid desaturation 39
  homeoviscous efficacy 8, 9
prolactin
  and mucus secretion 210
  and recognition of salinity changes 209
proline, insect fuel 97
protein
  changes and environmental stimuli 104
  polymorphism 122–4
  structure:function compromises 107
  turnover analysis 109
protein isoenzymes
  cathodic and cold acclimation 133
  composition and cold acclimation 131
  temperature-specific synthesis 134, 135
  thermal properties 129
protein isoforms 39, 124
  and temperature change 128, 129
Pseudopleuronectes americanus (winter flounder)
  antifreeze synthesis regulation 220, 221
  photoperiod and antifreeze 220
  seasonal antifreeze variation 219, 220
pump and leak concept 59, 60
pyruvate kinase in hibernating bats 129, 130

salinity
  adaptation in teleosts 197–210
  hormones and recognition of change 209, 210
  lipid polar head groups 44, 45
  mechanisms of adaptation 198
  recognition of change by fish 209, 210
Salmo gairdneri
  aflatoxin effects 188, 189, 194
  hepatocellular carcinoma and diet 188, 189, 194
  PCB induction of monooxygenase 184–7
  temperature acclimation and acetylcholinesterase properties 134, 135
Salvelinus fontinalis, ATPase and temperature 136, 137
sarcoplasmic reticulum, changes and cold acclimation 118
seawater
  adaptation changes in fish 198, 199
  gill changes and fish adaptation 202, 203
  osmolarity 197
sodium:chloride co-transport, activation trigger 75
sodium transport, temperature compensation and frog skin 23
Staphylococcus aureus, lipid polar head groups and salinity 44, 45
steroid hormones
  functions 191
  metabolism and monooxygenase induction 190, 191, 194
  PCB effects 190, 191
sterols, cell content and adaptation 46, 47
stress, environmental and fatty acid unsaturation 34, 35
succinate dehydrogenase 17, 27
surfentanyl 172
sycamore cell culture and fatty acid saturation 38

taurine and volume-regulatory decrease 69–71
temperature *see also* body temperature *and* cold
  ATP erythrocyte levels in fish 232, 233
  cytochrome c turnover 112–14
  enzyme properties 81
  fatty acid composition 43
  fatty acid synthesis 38
  fish blood pH 233
  fish oxygen requirement 231
  homeotherms, adaptation to cold 127
  lipid polar head groups 44, 45
  membrane adaptations 3–27
  and membrane fatty acids 3, 34
  mitochondria in goldfish 116, 117
  responses to seasonal changes 127–9
  subcellular enzyme concentrations 109–18
  weak bond stability 121
temperature compensation, metabolism in fish 109–14, 128
Tetrahymena sp.
  alcohol effects on membranes 148, 149

Tetrahymena sp. (*cont.*)
  chilling and fatty acid composition 42, 43
  lipid polar head groups and temperature 44, 45
  phase separation temperatures 12, 13
  temporal relationships and adaptation 47, 49
  time course of homeoviscous adaptation 10
tolerance 161
  opiate in guinea-pig ileum 163, 164
  receptor theory for opiates 168, 169
torpor
  diurnal 128
  seasonal 128
transport processes
  osmoregulatory 56
  and temperature 22, 23
tricarboxylic acid cycle (TCA) and muscle energy provision 88–96
triglyceride lipase 99
tropomyosin–troponin 138
  subunit forms 139

vas deferens
  naloxone effect 172
  opioid sensitivity 172, 173

volume regulation of cells 55–76
  anisotonic media 62–76
  cation permeability 65, 66
  Donnan equilibrium 57, 58
  mechanisms 56
  and medium 55
  ouabain effects 60
  potassium permeability 56
  pump and leak system 58–60
  sodium chloride co-transport 74, 75
volume-regulatory decrease 62–9
  amino acid role 69
  co-transport inhibitors 71, 73
  mechanisms 70
volume-regulatory increase 71–6
  mechanism 70
  nitrate ion effect 73

water and enzyme solubility 106
wheat epicotyl, fatty acid saturation 35
whelk radula muscle 89, 94
withdrawal, adaptive function 161

xenobiotic metabolism pathways 188
Xenopus laevis isoenzymes and cold 133

Yusho disease 190, 191